SpringerBriefs in Computer Science

T0213816

For further volumes:
http://www.springer.com/series/10028

Shibo He • Jiming Chen • Junkun Li • Youxian Sun

Energy-Efficient Area Coverage for Intruder Detection in Sensor Networks

 Springer

Shibo He
Department of Control
Science and Engineering
Zhejiang University
Hangzhou, Zhejiang
People's Republic of China

Junkun Li
Department of Control
Science and Engineering
Zhejiang University
Hangzhou, Zhejiang
People's Republic of China

Jiming Chen
Department of Control
Science and Engineering
Zhejiang University
Hangzhou
People's Republic of China

Youxian Sun
Department of Control
Science and Engineering
Zhejiang University
Hangzhou
People's Republic of China

ISSN 2191-5768   ISSN 2191-5776 (electronic)
ISBN 978-3-319-04647-1 ISBN 978-3-319-04648-8 (eBook)
DOI 10.1007/978-3-319-04648-8
Springer Cham Heidelberg New York Dordrecht London

Library of Congress Control Number: 2014931490

Printed on acid-free paper

Springer is part of Springer Science+Business Media (www.springer.com)

Preface

Due to the rapid development of wireless communication and hardware technologies, Wireless Sensor Networks (WSNs) are expected to be applicable in a large range of applications such as environmental surveillance (e.g., air pollution surveillance) and security monitoring (e.g., intrusion detection). One fundamental problem in WSNs is the coverage problem, which is about the placement and/or scheduling of sensors to maximize the quality of sensing in a deployment area, e.g., detect or capture interesting intruders. As technologies advance, practical sensors such as motes and smartdust may have small form factors and low cost. It is feasible to deploy a large number of these sensors for area monitoring. Carefully controlled placements of the sensors may be difficult, due to their large numbers or challenges of the geography. Instead, a loosely controlled method to place the sensors is more popular, which will result in random placements of the sensors. Moreover, unattended operation of the sensor network is often desirable or required. Therefore, there is a need to maximize the lifetime of the network before the sensors run out of energy, while the coverage performance is guaranteed. This problem is widely recognized as the *energy-efficient coverage* problem.

Traditionally, area coverage in WSNs requires that every point in the surveillance region be covered by the deployed sensor network. This definition can work for general applications where the surveillance region is of great interest, however, it is conservative and non-scalable. The coverage performance would not be impaired by a coverage hole as long as no interesting information is missed during the time of coverage hole. This also indicates that coverage performance hinges closely on the unique requirements of applications. There is thus a great potential to exploit the intrinsic characteristics of the applications to enhance both the energy-efficiency and coverage performance. An interesting and important application scenario is intruder detection, where a sensor network is deployed to detect stochastic intruders occurring at some points in the region. In such a scenario, different requirements of intruders capturing (e.g., detecting or tracing the intruders) can be leveraged to significantly enhance the performance.

In this book, we present some recent results on area coverage for intruder detection from an energy-efficient perspective. In Chap. 1, we first introduce the background, elaborate on system models such as the formal definition of area coverage and sensing models, and present a range of existing literatures and applications on area coverage. Then we focus on energy-efficient intruder detection under the well-known binary sensing model in Chap. 2, which specifically is devoted to showing how to improve the energy efficiency without impairing coverage performance by exploiting the dynamic nature of stochastic intruders. In Chap. 3, we proceed to investigate the intruder trapping under binary sensing model. Under such a scenario, any moving intruder will be detected by a sensor network when its movement distance is greater than a predefined constant. The network coverage holes act as traps. Once intruders fall into these traps, it can not get out without being detected. We design efficient algorithms to rotate the duty of each sensor to prolong the network lifetime while the intruder trapping performance is ensured. To investigate the impact of different sensing models, we investigate the intruder trapping under the probabilistic sensing model in Chap. 4. We conclude the book in Chap. 5.

Last but not least, we would like to thank Dr. Yu Gu, Dr. Tian He, Dr. Ten H. Lai, and Dr. David K.Y. Yau for their suggestions and contributions. We also would like to appreciate the Series Editor Dr. Xuemin (Sherman) Shen for his work and effort on this book.

Hangzhou, China   Shibo He
October, 2013   Jiming Chen
                                                                       Junkun Li
                                                                      Youxian Sun

Contents

1 Introduction to Area Coverage in Sensor Networks 1
  1.1 Background 1
  1.2 Basic Concepts 2
     1.2.1 Sensor Deployment 3
     1.2.2 Sensing Models 3
     1.2.3 Coverage Classification 4
  1.3 The State-of-the-Art Work on Area Coverage 5
     1.3.1 Deterministic Deployment 5
     1.3.2 Random Deployment 7
  References 8

2 Energy-Efficient Capture of Stochastic Events in Sensor Networks 11
  2.1 Introduction 11
  2.2 Problem Setup and Performance Metrics 13
  2.3 Event Capture by Periodic Sensor 15
  2.4 Energy-Aware Optimization of Synchronous Periodic Schedule 16
  2.5 Optimization of Asynchronous Periodic Schedule 18
  2.6 General Regionally Synchronous Networks 20
  2.7 Coordinated Sleep Under Periodic Scheduling 24
  2.8 Numerical Results 27
     2.8.1 Illustration of Analytical Results 27
     2.8.2 Network Simulations 29
     2.8.3 Summary of Experiments 32
  2.9 Conclusions 33
  References 34

3 Energy-Efficient Trap Coverage in Sensor Networks 35
  3.1 Introduction 35
  3.2 Preliminary and Problem Formulation 37
     3.2.1 Network Model 37
     3.2.2 Trap Coverage Model 38
     3.2.3 Minimum Weight Trap Cover Problem 39

3.3   Algorithm Design   40
       3.3.1   Finding the Diameter of a Coverage Hole   40
       3.3.2   Algorithm Overview   41
       3.3.3   Removal Strategy Design.......................................   42
       3.3.4   Algorithm Illustration   46
3.4   Performance Analysis...   47
       3.4.1   Theoretical Analysis   47
       3.4.2   Network Lifetime Analysis.....................................   50
       3.4.3   Simulation Performance   51
3.5   Localized Protocol   52
       3.5.1   Protocol   52
       3.5.2   Analysis   56
       3.5.3   How to Find the Largest Diameter   58
3.6   Simulation Results   60
       3.6.1   Experiment Setup   60
       3.6.2   Energy Balance and Consumption   60
       3.6.3   Lifetime Performance Evaluation   64
       3.6.4   Communication Cost   64
3.7   Conclusions   65
References   65

4   **Trapping Mobile Intruders in Sensor Networks**   69
4.1   Introduction   69
4.2   Preliminary and Problem Statement.....................................   72
       4.2.1   Network Model   72
       4.2.2   Probabilistic Trap Coverage Model   72
       4.2.3   Problem Statement..   73
4.3   Probabilistic Trap Coverage   74
       4.3.1   Detection Gain   74
       4.3.2   Impact of Maximum Speed   76
       4.3.3   Circular Graph   77
       4.3.4 (D, ϵ)-Trap Coverage..................................   82
       4.3.5   Solving an Open Problem in Barrier Coverage.................   83
4.4   Localized Protocol   84
       4.4.1   Probabilistic Trap Coverage Protocol..........................   84
       4.4.2   Protocol Analysis   87
4.5   Performance Evaluation   90
       4.5.1   Environment Setup   90
       4.5.2   Simulation Results   91
4.6   Conclusion   94
References   94

5   **Conclusions**   97

Chapter 1
Introduction to Area Coverage in Sensor Networks

1.1 Background

Integrated with the technologies of sensors, communication and Microelectro-mechanical systems (MEMS) [1], sensor networks have the great potential to bridge the gap between physical world and information world, and thus have been widely adopted in a large range of application domains, such as environmental surveillance (e.g., air pollution surveillance [2]) and security monitoring (e.g., intrusion detection [3]). At the core component of sensor networks is the functionality of sensing, i.e., collecting the important information from physical region of interest. This renders coverage problem one of the fundamental issues in sensor networks.

Generally speaking, coverage is an indicator of quality of sensing, quantifying to what extent we can know about the information in the region of interest via sensor networks [4]. It is thus a conceptual definition, and has different implications for varying application scenarios. Some specific coverage definitions have been introduced to characterize the specific requirements of applications. For example, *barrier coverage* was proposed to quantify the sensing requirements in applications where we need to detect intruders when they are crossing the region [3], and *point coverage* was presented to specify the quality of sensing in applications where we only focus on some discrete points of interest [5]. In this book, we concentrate on another interesting definition: *area coverage*, by which we are interested in information occurring in the whole region.

Obviously, the *area coverage* can be guaranteed if the deployed sensor network can retrieve the information at every point in the region. In view of this, traditionally every point in the region of interest is required to fall within the sensing region of at least one sensor so as to ensure *area coverage*. This approach, however, is conservative, and thus restricts the potential to improve the energy efficiency and *area coverage* performance. The coverage performance would not be impaired by a coverage hole as long as no interesting information is missed during the time of coverage hole. This indicates that the *area coverage* performance closely hinges on the unique requirements of applications. It is not desirable to have a generic

S. He et al., *Energy-Efficient Area Coverage for Intruder Detection in Sensor Networks*, SpringerBriefs in Computer Science, DOI 10.1007/978-3-319-04648-8_1,

coverage metric. In addition, from the perspective of network design, the choices of sensor distribution and sensing model have a great impact on the *area coverage* performance.

Sensors can be accurately installed at designated points to monitor the surveillance region. By optimizing sensor locations, sensors can form optimal network topology and thus improve the area coverage and reduce the total cost. Alternatively, a loosely controlled method to place the sensors may be used, which will result in random placement of the sensors. For example, an airplane is used to drop and scatter sensors to cover a mountainous terrain to detect the presence of animals and identify them. Moreover, unattended operation of the sensor network is often desirable or required. In such a scenario, there is a need to maximize the lifetime of the network before the sensors run out of energy, while the coverage performance is guaranteed. This problem is well recognized as the *energy efficient coverage*. To calculate the coverage of a region, the sensing model of sensors needs to be specified. The most popular model is the disc sensing model (a.k.a. binary sensing model), by which the sensing region of a sensor is a perfect disc. Due to its simplicity, the disc sensing model is widely adopted to derive theoretical results. Another frequently adopted model is probabilistic sensing model for signal-detection based sensors. With environmental noise, the signal received by sensors can be used to calculate the probability of the existence of an intruder. The probability is a decreasing function of the distance between a sensor and a point. The probabilistic sensing model is more practical than disc sensing model. The disadvantage is that it also brings high complexity to address the problem.

As technologies advance, practical sensors such as motes and smartdust may have small form factors and low costs. It is feasible to deploy a large number of these sensors for area monitoring. The large quantity of sensors and inaccessible geography incur huge cost for accurate installation. We thus in this book consider loosely controlled approach for sensor distribution, and focus on the challenging *energy efficient area coverage* problem. Most literatures addressed the problem in a conservative way, and did not consider the specific sensing requirements imposed by the applications. There is still a great potential to improve the *energy efficient area coverage*. Due to the diverse applicability of sensor networks, we concentrate on the scenario of intruder detection and demonstrate how to improve the energy efficiency by exploiting the unique characteristic of applications and the sensing models. These work may inspire new solutions to *energy efficient area coverage* in other application scenarios.

1.2 Basic Concepts

In this section, we introduce some basic concepts in the sensor networks, which will help understand the book.

1.2.1 Sensor Deployment

There are two approaches for sensor distribution in sensor networks:

Deterministic deployment. Given the deployed region, the optimal positions of sensor nodes can be pre-identified according to the chosen sensing model. Hence, the sensors can be placed deterministically to form the optimal topology. The most popular four patterns for sensor deployment are: (i) equilateral triangular, (ii) square, (iii) rhombus and (iv) hexagon [6]. Kershner [7] pointed out that the equilateral triangular pattern can obtain the minimum number of sensors to cover a plane, when the size of the plane is much larger than the sensing range of sensors, assuming that sensors have homogeneous disc sensing model. The square pattern is widely adopted due to its easy implementation. The rhombus and hexagon patterns may preserve desirable connectivity under certain conditions.

Random deployment. In the random deployment, sensors are scattered randomly by certain means (e.g., patrols or aircrafts), which may result in sensing redundancy and thus suboptimal network topology. It is typical that random deployment needs more sensors than deterministic deployment to achieve the same coverage performance. As the manufacture cost of a sensor has been inclining, this approach becomes prevailing due to its easy implementation. The random deployment process is widely modeled as uniform distribution or poisson point process [8].

1.2.2 Sensing Models

There are many sensing models proposed by researchers in the literature. We here introduce some popular ones.

Disc sensing model. This model assumes that the sensing region of a sensor is a circular region. When an event happens at the point within the circle centered at a sensor of radius r_s, this event is detected by the sensor [6]. Hereby, r_s is the sensing range. The disc sensing region is the inscribed circle of the real sensing region and thus is a rough approximation. It is widely adopted in the literature due to its simplicity and analyzability.

Probabilistic sensing model. Instead of deciding the coverage of a point deterministically such as disk sensing model, probabilistic sensing model assumes that the coverage function $f(x)$ is a probability function, ranging from 0 to 1 depending on the distance between the point and the sensor. Probabilistic sensing model is proposed for signal-detection based sensors. As there is environmental noise and signals decay during transmission, this model calculates the probability of the existence of an intruder. There are many probabilistic sensing models introduced in the literature. A popular one is defined as [9]

$$f(x) = \begin{cases} 1, & dist(x, P) \leq r_s \\ e^{-ka^m}, & r_s \leq dist(x, P) \leq r_u \\ 0, & r_u \leq dist(x, P) \end{cases} \tag{1.1}$$

where $a = dist(x, P) - r_s$, k and m are decay factors, and $r_u \gg r_s$ is a distance threshold dependent on the sensor hardware.

1.2.3 Coverage Classification

There are plenty of coverage definitions proposed by researchers. The three most popular ones are: (i) area coverage, (ii) point coverage, (iii) barrier coverage.

Area coverage. Given a 2-dimensional plane ω, area coverage is concerned with the coverage of the whole region ω. Traditionally, ω is covered if and only if every point in ω is covered by at least one sensor, and likewise it is k-covered when every point is covered by at least k sensors. *Area coverage* is the earliest definition introduced by researchers. When every point in the region is covered by a sensor network, there is definitely no information loss. *Area coverage* has been modified to work for 3-dimensional space or surface [10].

Point coverage. Some application scenarios may only concern with some discrete target points instead of the entire region. *Point coverage* is introduced to quantify the quality of sensing in such situations. Obviously, *point coverage* is easier to address than *area coverage*. It is shown in [5] that when the density of sensors is extremely high, each sensor can be viewed as a point and the corresponding *point coverage* will approximate closely the *area coverage*. Another interesting result is that after initial random deployment, the *energy-efficient area coverage* can be transformed to *energy-efficient point coverage* [11].

Barrier coverage. Unlike the area and point coverage, *barrier coverage* pays no attention to the events occurring in the deployed region [12,13]. Instead, it quantifies the probability that the intruders will be detected when they are attempting to cross the region [3]. By definition, *barrier coverage* does not require the entire even most of area to be covered. Therefore it is more scalable than *area coverage*. One of its promising applications is to detect the illegal intruders crossing the territory border.

  An illustration about these three definitions of coverage is plotted in Fig. 1.1a–c. Dependent on the sensing models and deployment approaches, the aforementioned definitions of coverage can be further refined. Specifically, *area coverage* can be further categorized as: (1) area coverage with disk sensing model under deterministic deployment, (2) area coverage with probabilistic sensing model under deterministic deployment, (3) area coverage with disk sensing model under random deployment and (4) area coverage with probabilistic sensing model under random deployment.

1.3   The State-of-the-Art Work on Area Coverage

There are many existing works on *area coverage* in the last decade.

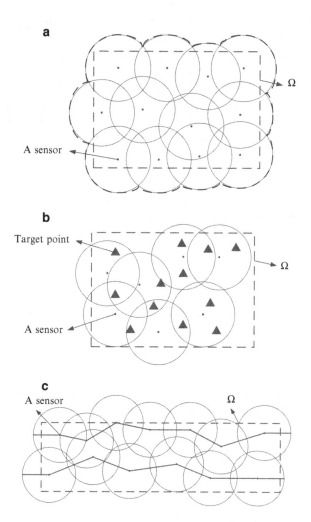

Fig. 1.1 An illustration of coverage. (**a**) Area coverage. (**b**) Point coverage. (**c**) Barrier coverage

1.3.1   Deterministic Deployment

Assuming that all sensors are homogeneous and have the same sensing range r_s, Kershner [7] firstly indicated that when the deployment region is a plane, whose length and width are much larger than the sensing range r_s, the equilateral triangle deployment pattern with side length of $\sqrt{3}r_s$ gives the minimum number of sensors to achieve full coverage. Zhang and Hou [14] obtained the same result from a different perspective. This finding addresses the 1-coverage problem and provides the benchmark for later coverage problem research. Based on the result, Bai et al. [15] considered 2-coverage, and showed that the deployment density should be at least $\frac{4\pi}{3\sqrt{3}}$ to achieve 2-coverage, and adopting equilateral triangle pattern twice (i.e., put two sensors at the same point) can optimally achieve 2-coverage with density of $\frac{4\pi}{3\sqrt{3}}$. Yu et al. [16] addressed the coverage problem with directional antennas. They adopted specifically two kinds of directional antenna models, and proposed optimal deployment patterns correspondingly. Moreover, they studied the robustness of the proposed algorithms, considering how do the decay and irregularity of practical sensing and communication models impact the performance of the algorithms.

It is essential to take connectivity into consideration when considering coverage [17]. To achieve coverage and connectivity at the same time, the benchmark result was given by Zhang and Hou [14]: when the communication range r_c is at least twice of the sensing range r_s, providing coverage implies connectivity; when $r_c \geq \sqrt{3}r_s$, equilateral triangle pattern aforementioned guarantees both coverage and connectivity. Bai et al. [6] investigated the optimal deployment for both coverage and connectivity when $r_c/r_s < \sqrt{3}$. They proposed a strip-based deployment pattern: the separation of neighboring sensors in the same strip is $\alpha = \min(r_c, \sqrt{r_s})$, and the vertical separation of neighboring strips is $\beta = r_s + \sqrt{r_s^2 - \alpha^2/4}$. They obtained two main conclusions: (1) in strip-based deployment, adding one sensor between every two neighboring strips can render the asymptotically optimal deployment to guarantee full-coverage and 1-connectivity; (2) adding two sensors between every two neighboring strips can render the asymptotically optimal deployment to achieve full-coverage and 2-connectivity. Based on [6], Bai et al. also considered full-coverage and 4-connectivity [18], and proposed rhombus deployment pattern, which divides the whole region into rhombuses, whose two diagonals, d_1 and d_2, satisfy

$$d_1 = 2r_s \cos\left(2\phi\sqrt{2(1 - \cos\phi)}\right)$$
$$d_2 = 2r_s \cos\left(2\phi\sqrt{2(1 + \cos\phi)}\right)$$

where $\phi = \max(2\arccos(r_c/2r_s))$, and sensors are placed on the vertexes of rhombuses. Bai et al. [18] pointed out that: (1) if $r_c/r_s \geq \sqrt{3}$, rhombus deployment pattern becomes triangular deployment; (2) if $r_c/r_s \leq \sqrt{2}$, rhombus deployment pattern is equivalent to square deployment pattern; (3) $r_c/r_s \geq \sqrt{2}$, rhombus deployment pattern is the optimal deployment to provide full-coverage and 4-connectivity. Bai et al. [19] also found that the hexagon deployment pattern is the optimal deployment when considering full-coverage and 3- and 5-connectivity. Bai et al. [15] investigated the problem of 2-coverage and 1- and 2-connectivity when $r_c > \frac{\sqrt{3}}{2}r_s$, as well as the problem of 2-coverage and 3-connectivity when $r_c > r_s$.

Compared with the 2-dimensional *area coverage*, 3-dimensional coverage problem is much more difficult. There are only a few work on this direction. Alam and Haas [10] demonstrated that Voronoi tessellation of 3-dimensional space to create truncated octahedral cells yields the optimal placement. In this deployment strategy, (i) when $r_c \geq 1.7889 r_s$, coverage can guarantee the connectivity, (ii) when $1.4142 \leq r_c/r_s \leq 1.7889$, the best placement is the hexagonal prism placement strategy or a rhombic dodecahedron placement. Bai et al. [20] considered full-coverage and k-connectivity ($k \leq 4$) at the same time, and proved the optimality of regular lattice deployment patterns under 1- and 2-connectivity. Furthermore, they considered more practical sensing and communication models for 3 dimensional coverage and connectivity, other than the sphere sensing and communication models.

1.3.2   Random Deployment

When sensors are initially deployed randomly with high density, an interesting problem is to duty-cycle sensors to improve *energy-efficient area coverage* [21–23]. Slijepcevic and Potkonjak designed a centralized offline algorithm [22], where they initially determined a group of sensor sets, and scheduled some subset of sensors work at each time to ensure k-coverage. Since offline algorithms fail to react properly to the dynamic environment such as the failure of sensors, it usually works as the theoretical benchmark to evaluate the performance of other distributed algorithms. Tian and Georganas [24] devised a distributed node-scheduling scheme, based on computational geometry, so as to determine if a sensor is redundant. Each sensor alternates between active and inactive states according to the result of redundancy check. Based on Tian and Georganas [24], Hsin and Liu [21] presented a distributed real-time algorithm, namely Role-Alternating Coverage-Preserving (RACP), to minimize sensors' sensing redundancy while maintaining coverage performance. They showed that coordinated sleep scheme outperforms random sleep scheme.

The coordinated sleep scheme can reduce the sensing redundancy significantly, however, there is still room to improve the result by global optimization of sleep scheme. Most of the existing work utilized the set cover theory to optimize the set selection [25–27]. Funke et al. [25] studied how to choose a connected coverage set with the minimum number of sensors. They first studied the greedy algorithm, and derived that the performance ratio of the result obtained by the greedy algorithm to that by the optimal is no more than $\log(m)$, where m is the number of sensors. Then, they proposed a grid based placement algorithm with a performance ratio 6π. Lastly, they presented a distributed dominating cover algorithm that has $\mathcal{O}(n)$ time complexity and $\mathcal{O}(n\log(n))$ message complexity and has a performance ratio 18. In [26], Misra and Mandal adopted minimum number connected dominating set to select the optimal set covers, and designed a collaborative cover heuristic, which obtains a partial Steiner tree under the construction of the independent set and outperforms the degree based heuristics. They obtained the performance ratio $(4.8 + \ln 5)opt + 1.2$, where opt is the size of any optimal connected dominating set. Wu and Li [27] investigated the problem of *k-Connected m-Dominating Set* (kmCDS) and developed a distributed local decision algorithm (LDA), which has $\mathcal{O}(n)$ message complexity and $\mathcal{O}(m + Diam)$ time complexity, where $Diam$ is the diameter of the network. For performance comparison, they additionally proposed a centralized algorithm, which has a constant approximation ratio.

The aforementioned work all adopted the disk sensing model. There are some works adopting other sensing models [28–30]. Jia et al. [29] adopted a probabilistic sensing model and proposed an elitist non-dominated sorting generic algorithm to select optimal cover set. Zou and Chakrabarty also [28] considered to deploy sensors for maintaining coverage with a probabilistic sensing model. They proved that the problem is NP-complete, and proposed a distributed algorithm based on dominating set theory, which can guarantee coverage and connectivity simultaneously. Moreover, Zhou et al. [30] considered variable radii *m-connected, k-coverage* problem in general sensor networks, proposed a distributed and localized Voronoi-based algorithm, which extends relative neighborhood graph (RNG) structure.

There are also many previous work on coverage hole detection under random deployment, which mainly are based on two main methods [31]: (i) radius-based [11, 32] and (ii) Voronoi diagram-based [33–35]. For the radius-based approach, the sensing boundary of a sensor is crucial for detecting coverage holes. Huang and Tseng [32] illustrated that the coverage holes occur if at least one point on the boundary of a sensor is not covered by its neighboring sensors. Moreover, Zhang and Hou [14] indicated that the coverage holes exist if the intersecting points of two sensors' sensing borders are not covered. Based on this finding, Kasbekar et al. [11] proposed a distributed algorithm without the knowledge of sensors' location information. For the Voronoi diagram-based method, Voronoi diagram is employed to detect the coverage holes. By utilizing local information, each sensor can independently decide whether any point in a Voronoi polygon is covered or not [33–35].

References

1. I. F. Akyildiz, W. Su, Y. Sankarasubramaniam, and E. Cayirci. Wireless sensor networks: A survey. *Computer Networks*, 38(4):393–422, 2002.
2. P. Dutta, P. M. Aoki, N. Kumar, and A. Mainwaring. Common sense: Participatory urban sensing using a network of handheld air quality monitors (demo abstract). In *Proceedings of the ACM Conference on Embedded Networked Sensor Systems (SenSys)*, 2009.
3. S. Kumar, T. Lai, and A. Arora. Barrier coverage with wireless sensors. In *Proceedings of the Annual International Conference on Mobile Computing and Networking (MobiCom)*, 2005.
4. S. Meguerdichian, F. Koushanfar, M. Potkonjak, and M. B. Srivastava. Coverage problems in wireless ad-hoc sensor networks. In *Proceedings of IEEE Conference on Computer Communications (INFOCOM)*, 2001.
5. S. Yang, F. Dai, M. Cardei, and J. Wu. On multiple point coverage in wireless sensor networks. In *Proceedings of IEEE International Conference on Mobile Adhoc and Sensor Systems Conference (MASS)*, 2005.

6. X. Bai, S. Kumar, D. Xuan, Z. Yun, and T. Lai. Deploying wireless sensors to achieve both coverage and connectivity. In *Proceedings of the ACM International Symposium on Mobile Ad Hoc Networking and Computing (MobiHoc)*, 2006.
7. R. Kershner. The number of circles covering a set. *American Journal of mathematics*, 61(3):665–671, 1939.
8. O. Dousse, P. Mannersalo, and P. Thiran. Latency of wireless sensor networks with uncoordinated power saving mechanisms. In *Proceedings of the ACM international symposium on Mobile ad hoc networking and computing (MobiHoc)*, 2004.
9. J. Chen, J. Li, S. He, Y. Sun, and H. Chen. Energy-efficient coverage based on probabilistic sensing model in wireless sensor networks. *IEEE Communication Letters*, 14(9):833–835, 2010.
10. S. M. N. Alam and Z. J. Haas. Coverage and connectivity in three-dimensional networks. In *Proceedings of the Annual International Conference on Mobile Computing and Networking (MobiCom)*, 2006.
11. G. Kasbekar, Y. Bejerano, and S. Sarkar. Lifetime and coverage guarantees through distributed coordinate-free sensor activation. In *Proceedings of the Annual International Conference on Mobile Computing and Networking (MobiCom)*, 2009.
12. S. He, J. Chen, X. Li, X. Shen, and Y. Sun. Cost-effective barrier coverage by mobile sensor networks. In *Proceedings of IEEE Conference on Computer Communications (INFOCOM)*, 2012.
13. S. He, X. Gong, J. Zhang, J. Chen, and Y. Sun. Barrier coverage in wireless sensor networks: From lined-based to curve-based deployment. In *Proceedings of IEEE Conference on Computer Communications (INFOCOM Mini Conference)*, 2013.
14. H. Zhang and J. C. Hou. Maintaining sensing coverage and connectivity in large sensor networks. *Journal of Wireless Ad-hoc and Sensor Networks*, 1(3):89–124, 2005.
15. X. Bai, Z. Yun, D. Xuan, B. Chen, and W. Zhao. Optimal multiple-coverage of sensor networks. In *Proceedings of IEEE Conference on Computer Communications (INFOCOM)*, 2011.
16. Z. Yu, J. Teng, X. Bai, D. Xuan, and W. Jia. Connected coverage in wireless networks with directional antennas. In *Proceedings of IEEE Conference on Computer Communications (INFOCOM)*, 2011.
17. C. Bettstetter and C. Hartmann. Connectivity of wireless multihop networks in a shadow fading environment. *Wireless Networks*, 11(5):571–579, 2005.
18. X. Bai, Z. Yun, D. Xuan, T. H. Lai, and W. Jia. Deploying four-connectivity and full-coverage wireless sensor networks. In *Proceedings of IEEE Conference on Computer Communications (INFOCOM)*, 2008.
19. X. Bai, D. Xuan, Z. Yun, T. H. Lai, and W. Jia. Complete optimal deployment patterns for full-coverage and k-connectivity ($k \leq 6$) wireless sensor networks. In *Proceedings of ACM International Symposium on Mobile Ad Hoc Networking and Computing (MobiHoc)*, 2008.
20. X. Bai, C. Zhang, D. Xuan, J. Teng, and W. Jia. Low-connectivity and full-coverage three dimensional wireless sensor networks. In *Proceedings of ACM International Symposium on Mobile Ad Hoc Networking and Computing (MobiHoc)*, 2009.
21. C. Hsin and M. Liu. Network coverage using low duty-cycled sensors: random and coordinated sleep algorithms. In *Proceedings of ACM/IEEE International Conference on Information Processing in Sensor Networks (IPSN)*, 2004.
22. S. Slijepcevic and M. Potkonjak. Power efficient organization of wireless sensor networks. In *IEEE International Conference on Communications (ICC)*, pages 472–476, 2001.
23. C. Chiasserini and M. Garetto. An analytical model for wireless sensor networks with sleeping nodes. *IEEE Transactions on Mobile Computing*, 5(12):1706–1718, 2006.
24. D. Tian and N. Georganas. A coverage preserving node scheduling scheme for large wireless sensor networks. In *Proceedings of ACM International Workshop on Wireless Sensor Networks and Applications (WSNA)*, 2002.
25. S. Funke, A. Kesselman, F. Kuhn, Z. Lotker, and M. Segal. Improved approximation algorithms for connected sensor cover. *Wireless Networks*, 13(2):153–164, 2007.

26. R. Misra and C. Mandal. Minimum connected dominating set using a collaborative cover heuristic for ad hoc sensor networks. *IEEE Transactions on Parallel and Distributed Systems*, 21(3):292–302, 2010.

27. Y. Wu and Y. Li. Construction algorithms for k-connected m-dominating sets in wireless sensor networks. In *Proceedings of ACM International Symposium on Mobile Ad Hoc Networking and Computing (MobiHoc)*, 2008.

28. Y. Zou and K. Chakrabarty. Sensor deployment and target localization in distributed sensor networks. *ACM Transactions on Embedded Computing Systems*, 3(1):61–91, 2004.

29. J. Jia, J. Chen, G. Chang, and Y. Wen. Efficient cover set selection in wireless sensor networks. *Acta Automatica Sinica*, 34(9):1157–1162, 2008.

30. Z. Zhou, S. Das, and H. Gupta. Fault tolerant connected sensor cover with variable sensing and transmission ranges. In *Proceedings of IEEE Annual Conference on Sensor and Ad Hoc Communications and Networks (SECON)*, 2005.

31. C. Zhang, Y. Zhang, and Y. Fang. Localized algorithms for coverage boundary detection in wireless sensor networks. *Wireless Network*, 15(1):3–20, 2009.

32. C. Huang and Y. Tseng. The coverage problem in a wireless sensor network. In *Proceedings of ACM International Workshop on Wireless Sensor Networks and Applications (WSNA)*, 2003.

33. Q. Fang, J. Gao, and L. Guibas. Locating and bypassing routing holes in sensor networks. In *Proceedings of IEEE Conference on Computer Communications (INFOCOM)*, 2004.

34. A. Ghosh. Estimating coverage holes and enhancing coverage in mixed sensor networks. In *Proceedings of IEEE Annual Conference on Local Computer Networks (LCN)*, 2004.

35. S. Kapoor and X. Li. Proximity structures for geometric graphs. In *Proceedings of Workshop on Algorithm and Data Structure (WADS)*, 2003.

Chapter 2
Energy-Efficient Capture of Stochastic Events in Sensor Networks

2.1 Introduction

In this chapter, we consider the *energy-efficient coverage* in a sensor network with an initial random sensor deployment, aiming at capturing stochastic events occurring in the surveillance region [1]. To achieve the goal of energy efficiency, there is a need to duty-cycle the sensors to minimize the redundancy of coverage caused by overlap in their sensing regions. In a *coordinated sleep* approach, we determine when a sensor, is made redundant because its sensing region is completely covered by those of its active neighbors. We can then safely turn off the sensor to conserve energy without hurting the performance. Moreover, it is desirable to rotate the active sensors to achieve energy balance, so that different subsets of the sensors are active at different times. The load balancing prolongs the lifetime of the network before a significant number of the sensors die and cause a severe loss in the coverage [2].

Prior work in coordinated sleep takes a conservative approach. It tries to ensure that every point of the deployment area is covered all the time by at least one active sensor, provided that there is an available sensor in the random placement. The *Role-Alternating*, Coverage-Preserving (RACP) algorithm in [3] is designed to minimize the probability that any given point is not covered by an active sensor, provided that it could be covered. Doing so will also maximize the detection probability of any event and ensure instantaneous detection, if the event is within range of at least one sensor that is alive. In some applications for transient events (e.g., an animal which arrives and then leaves), the goal may be to maximize the detection probability of the events before they disappear, and some delay in the detection is acceptable. Given the relaxed performance objective, leaving an area uncovered some of the time may be acceptable, since an event arriving when there is no active sensor may stay long enough until a sensor becomes active. For simplicity, we can consider periodic scheduling of the individual sensors, in which a sensor is active for only q time every p time ($q \leq p$). It has been shown in [4,5] that knowledge about the event dynamics—the stochastic processes of the event arrivals/departures—can be used to optimize the periodic schedule of a single sensor for event capture.

S. He et al., *Energy-Efficient Area Coverage for Intruder Detection in Sensor Networks*,   11
SpringerBriefs in Computer Science, DOI 10.1007/978-3-319-04648-8_2,
© The Author(s) 2014

Our main contribution in this chapter is to expose the interactions between periodic scheduling and coordinated sleep in a dense static sensor network. We focus on periodic schedules because they can be easily implemented and have been extensively employed with success in diverse settings; in particular, they already allow a wide and systematic spectrum of performance and cost tradeoffs [6]. We will extend the results in [4, 5] to consider energy efficiency and the collective performance of the sensors. The basic observation is that if events may stay for some time, then for a point covered periodically for q time every p time, the fraction of events captured there may be significantly more than q/p. Given the (q, p) schedules of the sensors, the global network itself will achieve the same periodic coverage schedule if the on periods of the sensors are *synchronous* (i.e., they start at the same time). If the on period of each sensor starts at a uniformly random point within p, then the periodic sensor schedules are *asynchronous*. We derive the optimal periodic schedule (q^*, p^*) for both kinds of network. More generally, the size of the sensor synchronization region is specifiable, which gives rise to a spectrum of *regionally synchronous* networks between the synchronous and asynchronous networks. Coordinated sleep by the sensors can then be applied orthogonally to further improve the energy efficiency. We consider the interactions between the periodic scheduling and coordinated sleep, including the cases of (i) synchronous periodic coverage without coordinated sleep (S-nc), (ii) synchronous periodic coverage with coordinated sleep (S-CSP), (iii) asynchronous periodic coverage without coordinated sleep (A-nc), and (iv) asynchronous periodic coverage with coordinated sleep (A-CSP).

Apart from its simplicity, S-nc is mostly interesting as a basis for performance comparison, since its performance is otherwise dominated by that of S-CSP. All the other approaches of S-CSP, A-nc, and A-CSP are of practical interest, and it is instructive to compare their performance with each other and with the RACP protocol in [3]. We have the following findings.

(i) Among the periodic scheduling approaches, S-CSP maximizes the opportunities for coordinated sleep. It can thus achieve the longest network lifetime at the price of some performance loss in event capture. When q is small compared with p, S-CSP is significantly more energy-efficient than RACP, but it is inferior to RACP in terms of capturing events without delay. The performance gap closes significantly, however, when we measure the fraction of events captured with or without delay. S-nc performs better in this measure because it is designed to exploit the possibility of capturing an event before the event leaves.

(ii) A-nc has the least overhead among all the approaches because it requires zero sensor coordination. It has the same network lifetime as S-nc (hence shorter lifetime than S-CSP) but it has better event-capture performance than the synchronous approaches, by spreading out the redundant coverage for a higher global coverage intensity. Under a wide range of q/p, its performance in terms of event capture (with or without delay) is extremely competitive with that of RACP and it has a longer network lifetime than RACP. When the sensor density is high, it is extremely competitive with RACP in terms of *instantaneous* event capture also, while remaining more energy efficient than RACP. Hence, periodically resting the sensors to conserve energy does *not* necessarily add noticeable delay in the event capture. Moreover, this extremely high performance for *instantaneous* event capture is by nature independent of the event dynamics.

(iii) A-CSP achieves the same event-capture performance (instantaneous or delayed) as A-nc (hence the same competitive performance with RACP), but it can further extend the network lifetime beyond A-nc. When the sensors' periodic schedules are asynchronous, however, the chance for coordinated sleep decreases. A higher sensor density will then help A-CSP realize its potential for energy savings, provided *also* that q is large enough relative to the energy costs of turning off/on the sensor. More generally, as the size of the synchronization region decreases, the event capture performance increases while the opportunities for coordinated sleep decrease. Our performance results show that *if the sensor density is high and both instantaneous and delayed capture performance are important, A-CSP represents the overall best method.*

The rest of the chapter is organized as follows. We formulate the problem of energy-efficient event capture in Sect. 2.2. We present results on event capture by a periodic sensor in Sect. 2.3, based on which the optimization of synchronous and asynchronous periodic schedules is derived in Sects. 2.4 and 2.5, respectively. We propose a general paradigm of regionally synchronous networks in Sect. 2.6. A coordinated sleep protocol under periodic scheduling is presented in Sect. 2.7. We present extensive simulation results in Sect. 2.8 for performance evaluation, and conclude in Sect. 2.9.

2.2 Problem Setup and Performance Metrics

We assume that sensors are used to cover an area for capturing stochastic events. We use the perfect disc sensing model, meaning that an event is captured if its distance from an active sensor is less than distance r, where r is the sensing range. The perfect disc sensing model is widely used in the literature due to its simplicity [7, 8]. While more complex anisotropic models [9] could be used for increased accuracy, doing so will not change our conclusions qualitatively. We assume that sensors can be turned off (made inactive) independently to conserve energy. Strictly speaking, each sensor has three main functional modules—sensing, communication, and computation—which can be turned off independently. For simplicity, we will assume that when the sensor becomes inactive, all three functional modules are inactive. According to specifications of real-world sensors such as Crossbow motes [10], the sensor in the active and inactive states consumes energy at rates of k_1 and k_2, respectively, where k_1 and k_2 are constants. Moreover, a constant amount of energy, given by c, is needed for the sensor to change between the active/inactive states.

We assume that transient random events appear at given *points of interest* (PoIs). The average rate of event arrivals at a PoI is given by λ. After an event arrives at PoI i, it stays for a random *event staying time* drawn from a distribution X. After that, the event disappears and after another random *event absent time* drawn from a distribution Y, the next event arrives at i. We refer to the statistical characteristics of the event staying and absent times as the *event dynamics*. We assume that the staying time and the subsequent event absent time of different events are i.i.d., even though the two quantities for the same event may be dependent. We do not exploit the spatial correlation of PoIs in the network design. Hence, we assume that the event dynamics at different PoIs are independent. This model of event dynamics applies to many application scenarios in event detection. One example is the characterization of event dynamics in mobile sensor networks [4]. Another example is spectrum detection in cognitive sensor networks [11]. A primary user may occupy a channel for some time period, after which the primary user may leave the channel, and so on.

We consider random placement of the sensors according to a Poisson point process of intensity γ. The Poisson point process is widely used in the analysis of random node placements in networks [12]. When γ is large, there is significant redundancy between the sensing regions of sensors. A sensor S is made redundant by its active neighbors if its sensing region is completely covered by those of the neighbors. Moreover, we assume that there is a uniform γ for the whole network for simplicity. If the sensor density varies, our analysis can be applied to each of the varying parts in a straightforward manner.

In quantifying the performance of the sensor network, we restrict our attention *only* to PoIs that are within distance r of at least one sensor. It is because if events happen at PoIs not within range of any sensor, the events cannot be captured irrespective of the network design,[1] and we consider these events out of scope. We use two principal performance metrics: (i) the probability of instantaneous capture, P_{in}, which is the probability that an event arriving at a PoI will be captured immediately upon arrival (without delay); and (ii) the probability of capture, P_c, which is the probability that an event appearing in a PoI will be captured before it leaves the PoI (though the event does not necessarily leave the network). It is clear that $P_c \geq P_{in}$. In many applications, some delay in the detection is acceptable. In a wildlife monitoring network, for example, researchers may be interested in identifying the animals that pass by a given point, but they do not need immediate report of the information. Also, when events are detected with delays, the average delay can be further quantified for performance evaluation.

---

[1]Since we do not control the sensor placement.

2.3 Event Capture by Periodic Sensor

We now analyze the per-PoI event capture performance of a sensor on a periodic schedule (q, p). The periodic schedule means that before energy runs out, the sensor is alternately active and inactive for q time and $p - q$ time, respectively. The following theorem concerns the probability that an event is captured by the sensor and is a paraphrase of Theorem 2 in [5].

Theorem 2.1. *For a (q, p) periodic sensor whose sensing region covers PoI i, the sensor captures an event at i (before the event disappears) with probability*

$$P_c = \frac{q}{p} + \frac{1}{p} \int_0^{p-q} \Pr(X \geq t)\, dt$$

before it runs out of energy.

Proof. Refer to [5], Theorem 2.

Note that the expression for P_c is a sum of two terms. The first term gives the fraction of events captured during the sensor present period $[0, q]$. It is equivalent to the performance measure P_{in} since these events are captured instantaneously. The second term gives the fraction of captured events that arrive during the sensor absent period $[q, p]$, but stay long enough to be captured during the *next* sensor present period $[p, p+q]$. These events are captured with a delay, and therefore contribute to the performance measure P_c. The main observation is that due to the contribution of the second term, the sensor working for q time every p time may capture a fraction of events that is much higher than q/p. This property holds true as long as the events stay, as they typically do in real applications, and does not depend on the detailed event dynamics. We intend to exploit this property of the periodic sensor to achieve high energy savings with relatively minor loss in the capture performance.

We can analyze the expected delay for those events captured with a delay, as well as for all the captured events, as given by the following theorem.

Theorem 2.2. *For the events captured at PoI i that arrive during a sensor absent period, the expected delay until their capture at i is given by*

$$D_{del} = \frac{\int_0^{p-q} \Pr(X \geq t) \times t\, dt}{\int_0^{p-q} \Pr(X \geq t)\, dt}.$$

For all the events captured at the PoI i, the expected capture delay is given by

$$D_{tot} = \frac{\int_0^{p-q} \Pr(X \geq t) \times t\, dt}{q + \int_0^{p-q} \Pr(X \geq t)\, dt}.$$

Proof. An event arriving at t time before the next sensor present period will be captured with delay t provided that it stays long enough. Since the event must arrive during the current sensor absent period, the value of t ranges from 0 to $p - q$, and $\int_0^{p-q} \Pr(X \geq t) \, dt$ is the probability that the event stays long enough. Therefore, $\int_0^{p-q} \Pr(X \geq t) \times t \, dt$ is the average delay of all the captured events. $\int_0^{p-q} \Pr(X \geq t) \, dt$ expresses the probability that an event occurs during the absent period. Hence, $\frac{\int_0^{p-q} \Pr(X \geq t) \times t \, dt}{\int_0^{p-q} \Pr(X \geq t) \, dt}$ gives the expected capture delay of the events that arrive during the absent period. A similar computation yields the value of D_{tot}.

Theorem 2.1 applies to general distributions of the event staying times. We can illustrate the result using the Exponential distribution with rate parameter α, i.e., the pdf of X, denoted by $f(x)$, is given by:

$$f(x) = \alpha e^{-\alpha x}, \quad x > 0, \quad \text{mean} = \frac{1}{\alpha}.$$

Then we have

$$P_c[X \in \text{Exponential}(\alpha)] = \frac{q}{p} + \frac{1 - e^{-\alpha(p-q)}}{p\alpha}. \tag{2.1}$$

Similarly, the delay values in Theorem 2.2 specialize to

$$D_{del}[X \in \text{Exponential}(\alpha)] = \frac{1 - e^{-\alpha s}(\alpha s + 1)}{\alpha(1 - e^{-\alpha s})}, \tag{2.2}$$

$$D_{tot}[X \in \text{Exponential}(\alpha)] = \frac{1 - e^{-\alpha s}(\alpha s + 1)}{\alpha(\alpha q + 1 - e^{-\alpha s})}, \tag{2.3}$$

where $s = p - q$.

2.4 Energy-Aware Optimization of Synchronous Periodic Schedule

We discuss optimization of the periodic schedule (q, p) for the *synchronous* network. In such a network, all the sensors employ the same (q, p) schedule, and they start their on periods at the same time so that the on/off periods are synchronized. Thus, the global network as a whole behaves like one big sensor on the (q, p)-periodic schedule. In practice, lightweight and accurate time synchronization protocols are available [13] to support the implementation of the synchronous network.

It can be shown that, for general distributions of X, the P_c value in Theorem 2.1 has two properties: (i) for the same p, P_c is monotonically increasing in q/p; (ii) for the same q/p, P_c is monotonically decreasing in p. Hence, to optimize the event capture, we can either make $\frac{q}{p} = 1$, or if we decide to use a smaller q/p, we can make p as small as possible. The first option will not help us save energy, while the second option is limited physically by the delay and energy expense in switching the sensor between the on/off states frequently.

To optimize the periodic schedule, we need to explicitly account for energy use by the energy model given in Sect. 2.2.[2] Specifically, we know that for a (q, p)-periodic sensor, it is able to capture a fraction P_c of events according to Theorem 2.1. Hence, its per-PoI rate of capturing events is given by $Q' = \lambda \times P_c$. On the other hand, the rate of energy use of such a sensor is given by

$$E' = \frac{k_1 \cdot q + k_2(p - q) + 2c}{p}. \tag{2.4}$$

We allow the user to specify the minimum P_{in}, denoted by P_{in}^{min}, that events are detected without delay, where $P_{in}^{min} > 0$. By Theorem 2.1, P_{in} is equal to q/p. Hence, we set $\frac{q^*}{p^*} = P_{in}^{min}$ and optimize P_c by solving the following optimization problem for the *per-PoI number of events captured per unit of energy*, denoted by Q_E:

$$\text{Find } p^* = \arg\max_p \ Q_E = \arg\max_p \ Q'/E'.$$

(Note that the above formulation is equivalent to optimizing the expected number of events captured before an energy budget given by B is depleted.) For example, if $X \in \text{Exponential}(\alpha)$ and $Y \in \text{Exponential}(\beta)$, then $\lambda = \alpha\beta/(\alpha + \beta)$. By (2.1), the rate of capturing events at a PoI is

$$Q' = \frac{\beta}{p(\alpha + \beta)}(1 + q\alpha - e^{-\alpha(p-q)}). \tag{2.5}$$

The optimization is then to find p^* that maximizes

$$Q_E = \frac{\beta}{\alpha + \beta} \frac{1 + q\alpha - e^{-\alpha(p-q)}}{k_1 \cdot q + k_2(p - q) + 2c}, \tag{2.6}$$

where $q = P_{in}^{min} \times p$. Note that we have a one-dimensional optimization problem, and the solution can be computed numerically by comparing the points of p where

$$\frac{dQ_E}{dp} = 0 \text{ and } \frac{d^2Q_E}{dp^2} < 0.$$

---

[2]For simplicity, we do not consider the latency constraints of turning on/off the sensor, but note that such constraints can be easily incorporated.

Remark. Note that in the network, more than one PoI may be within range of a sensor, and the same PoI may be within range of more than one sensor. Hence, the sensors may not capture events that are distinct. Assume that there are m distinct PoIs within range of n sensors in the network. The number of distinct events captured per unit energy for the whole network is $\frac{m \times Q'}{n \times E'}$. This value is the Q_E value derived above scaled by a constant factor of $\frac{m}{n}$. The scaling by a constant factor will *not* affect the solutions to the optimization problem.

2.5   Optimization of Asynchronous Periodic Schedule

We now analyze the *asynchronous* network. In this kind of network, each sensor employs the same (q, p)-periodic schedule, but they start their on periods independently at a uniformly random point in time within the period p. Because the on periods of the sensors are spread out in the asynchronous network, the event capture performance must consider the joint operation of the sensors. We have the following main results (Theorems 2.3 and 2.4).

Theorem 2.3. *For a random placement of sensors by the Poisson point process of intensity γ, the probability that an event appearing at a PoI is captured instantaneously is given by $P_{in} = \frac{1-e^{-\gamma \pi r^2 \frac{q}{p}}}{1-e^{-\gamma \pi r^2}}$, where r is the sensing range.*

Proof. A sensor can potentially detect an event if the event happens within distance r of the sensor. Hence, an event may be potentially detected by any sensor within the circular region centered at the event and of radius r. Note that at least one such sensor exists, by the definitions of our performance metrics P_{in} and P_c. By the property of the Poisson point process, the probability p_k that there are k sensors in the circular region is given by

$$p_k = \frac{(\gamma \pi r^2)^k e^{-\gamma \pi r^2}}{k!}. \tag{2.7}$$

The probability that any sensor is inactive when the event happens is $1 - \frac{q}{p}$. The event is undetected on arrival only if all the k sensors are inactive, which happens with probability $(1 - \frac{q}{p})^k$. Hence, summing over the range of k, we can compute the probability that the event is undetected on arrival as

$$\frac{1}{1 - p_0} \sum_{k=1}^{\infty} (1 - \frac{q}{p})^k \, p_k = \frac{e^{-\gamma \pi r^2 \frac{q}{p}} - p_0}{1 - p_0}.$$

The result follows as the complement of the above probability.

Remark. Whereas $1 - P_{in}$ decreases linearly with $\frac{q}{p}$ in the synchronous network, the decrease is exponential in the asynchronous case. This implies that for the asynchronous network, a small increase in $\frac{q}{p}$ may result in a large increase in P_{in}. We will show in Sect. 2.8.2 that the asynchronous network can indeed achieve high energy efficiency with an extremely small loss in event capture performance. Also, when the network is configured to achieve an extremely high instantaneous capture probability, its excellent performance is in fact *not* dependent on the event dynamics.

Fig. 2.1 Illustration of the synchronous, regionally synchronous, and asynchronous networks. (**a**) Synchronous network. (**b**) Regionally synchronous network. (**c**) Asynchronous network

Theorem 2.4 ($X \in$ Exponential(α)). *For a random placement of sensors by the Poisson point process of intensity γ, the probability that an event appearing at a PoI is captured before it leaves the PoI is given by $P_c = \frac{1 - e^{\gamma \pi r^2 (\rho - 1)}}{1 - e^{-\gamma \pi r^2}}$, where r is the sensing range and $\rho = \frac{\alpha(p-q) - 1 + e^{-\alpha(p-q)}}{p \times \alpha}$.*

Proof. Let $s = p - q$. Consider a sensor in the circular region of radius r centered at the event. The sensor starts its most recent off period at time 0. Consider an event arriving at time t and staying for time x, where $x \sim X \in$ Exponential(α). Without loss of generality, we assume that the start time t of the event is within the period $[0, p]$. The sensor does not detect the event if it is inactive in $[t, t+x]$, which occurs with probability $\Pr(x < s - t)$ provided that $t < s$. Hence, summing over the range of t and denoting by ρ the probability that the sensor does not detect the event before the event leaves, we have

$$\rho = \frac{1}{p} \int_0^s 1 - e^{-\alpha(s-t)} \, dt = \frac{\alpha s - 1 + e^{-\alpha s}}{p \times \alpha}.$$

For the Poisson point process, (2.7) gives the probability that there are p_k sensors within range of the event. The probability that the event is undetected by any in-range sensor before it leaves is therefore

$$\frac{1}{1 - p_0} \sum_{k=1}^{\infty} p_k \, \rho^k = \frac{e^{\gamma \pi r^2 (\rho - 1)} - p_0}{1 - p_0}.$$

The result follows as the complement of the above probability.

According to Theorem 2.3, P_{in} does not depend on the event dynamics X and Y. From Theorem 2.4, P_c only depends on the event staying time X, which is only used to calculate ρ. Similar results can be obtained for other distributions of X. For ease of exposition, we will present results for Exponential event staying times only.

As with the synchronous network, we allow the user to specify the minimum probability $P_{in}^{\min}$ of instantaneous event detection. Notice that P_{in} is monotonically increasing in both the sensor density γ and the ratio q/p. When $q = p$, $P_{in} = 1$. Hence, any P_{in} is satisfiable. We can compute the smallest q/p needed to satisfy $P_{in}^{\min}$ and denote this value by z. Once z is determined, we can find p^* (hence $q^* = z\,p^*$) as the optimal p that maximizes $Q_E = \frac{P_c \times \lambda}{E'}$. As in the synchronous network case, we have a one-dimensional optimization problem, although the expression for P_c is more complex.

Note also that if we considered the capture probabilities of *all* events (i.e., whether they fall within range of at least one sensor or not), the above Q_E measure would be scaled by $1 - p_0$, which is independent of p and q and hence will not affect the optimization problem.

2.6 General Regionally Synchronous Networks

We have discussed the design of synchronous and asynchronous networks. In a synchronous network, the sensors are network-wide synchronized, i.e., they begin their on periods at the same time (Fig. 2.1a). In an asynchronous network, the sensors do not synchronize but start their on periods independently (see Fig. 2.1c). The asynchronous network outperforms the synchronous network in terms of P_{in} and P_c. However, when neighboring sensors may also coordinate to eliminate their coverage redundancy (the design of the supporting coordinated sleep protocol is the subject of Sect. 2.7), the synchronous network may maximize the coordinated sleep opportunities because the synchronized on periods of neighboring sensors will better allow these sensors to cover for each other. Hence, sensor synchronization may have an advantage in energy efficiency.

In view of the energy saving potential of synchronous networks, we now generalize the synchronous network into a class of *regionally synchronized networks* in which the network is partitioned into a number of disjoint synchronization regions and the size of each region is specifiable. Specifically, we divide the network into a specifiable number of K equal size *clusters*. Each cluster defines a synchronization region in that all the sensors within the same cluster synchronize their periodic on/off schedules. However, each cluster will choose the beginning of its on period randomly and independently of the other clusters. Hence, although sensors within the same cluster are synchronous, different clusters are asynchronous with respect to each other. Figure 2.1b illustrates the regionally synchronized network. We assume that each sensor will belong to one and only one cluster, and it is clear that the larger the value of K, the smaller is the size of the cluster or synchronization region.

Regionally synchronous networks are interesting because they may reduce the synchronization overhead of the (global) synchronous network in that synchronization messages from a sensor will not need to be disseminated throughout the network but only within a local region. This is important because sensors are generally small devices with constrained computation and communication capabilities. Moreover, given that the (global) asynchronous network is expected to be more effective in event capture but less effective in coordinated sleep than the (global) synchronous network, the regionally synchronous network paradigm provides the intermediate design points so that a broad spectrum of tradeoffs between the event capture and coordinated sleep performance can be enabled.

We proceed to analyze the performance of the regionally synchronous network. We assume that the network is divided into K disjoint equal size clusters. For an arbitrary PoI x in the network, the circular region centered at x of radius r is then divided into several sub-regions depending on how many clusters intersect the circular region. Assume that cluster i has an intersection of size s_i with the circular region; clearly, s_i is a function of x and ranges within $[0, \pi r^2]$, and $\sum_{i=1}^{K} s_i(x) = \pi r^2$. If cluster i does not intersect the circular region, $s_i = 0$. Conversely, if the circular region is fully contained in cluster i, $s_i = \pi r^2$. For an event occurring at x, we can analyze the probabilities $P_{in}(x)$ and $P_c(x)$ as follows.

Lemma 2.1 ($X \in$ **Exponential**(α)). *Assume the network has K clusters, for a specific PoI x, the probabilities $P_{in}^K(x)$ and $P_c^K(x)$ of capturing an event occurring at x instantaneously and before the event disappears, respectively, are given by*

$$P_{in}^K(x) = \frac{1 - \prod_{i=1}^{K}(1 - \frac{q}{p}(1 - e^{-\gamma s_i(x)}))}{1 - e^{-\gamma \pi r^2}}, \tag{2.8}$$

$$P_c^K(x) = \frac{1 - \prod_{i=1}^{K}(1 - (1 - p)(1 - e^{-\gamma s_i(x)}))}{1 - e^{-\gamma \pi r^2}}. \tag{2.9}$$

Proof. Recall $\frac{q}{p}$ and ρ in Theorem 2.4, which are denoted as the probabilities of capturing an event instantaneously and before it disappears, respectively. In each cluster, all the sensors behave like one big periodic sensor. For cluster i and PoI x, i has an area of size $s_i(x)$ intersecting with the circle centered at x of radius r. The probabilities that it can capture the event instantaneously or before it disappears are equal to $\frac{q}{p}(1 - e^{-\gamma s_i(x)})$ and $(1 - p)(1 - e^{-\gamma s_i(x)})$, respectively. The event is captured if at least one of the clusters captures it, and the result follows.

As K changes, $P_{in}^K(x)$ and $P_c^K(x)$ will change. Even for the same K, the clustering can be performed in many different ways leading to different values of $s_i(x)$. For example, for the three clusters shown in Fig. 2.1b, $s_i(x_1)$ and $s_i(x_2)$ are different. $P_{in}^3(x_1)$ and $P_c^3(x_1)$ are the same as the values for the (global) synchronous network because x_1 is fully contained in one of the clusters. $P_{in}^3(x_2)$

and $P_c^3(x_2)$ are larger than the corresponding values for x_1, since x_2 lies near the boundary of the clusters so that it can be covered by more than one clusters. Hence, the values of $P_{in}^K(x)$ and $P_c^K(x)$ are dependent on the location of x. In general, the values of P_{in} and P_c for the whole regionally synchronous network will be higher than the corresponding values for the (global) synchronous network (i.e., $K = 1$) but smaller than those of the asynchronous network. When $K \to \infty$ and $\max A_i \to 0$, where A_i is the area of the cluster i, the regionally synchronous network should become like the asynchronous network. The following theorem formalizes this intuition.

Theorem 2.5 ($X \in$ **Exponential**(α)). *For general K, the network performance of the regionally synchronous network is less good than the asynchronous network but better than the synchronized network, i.e., $\forall K$, we have $q/p \leq P_{in}^K(x) \leq P_{in}^{async}$ and $1 - \rho \leq P_c^K(x) \leq P_c^{async}$, where P_{in}^{async} and P_c^{async} are the P_{in} and P_c of the asynchronous network. Furthermore, when the number of clusters $K \to \infty$ and $\max A_i \to 0$, the regionally synchronous network performs the same as the asynchronous network in terms of P_{in} and P_c,[3] i.e.,*

$$\lim_{K \to \infty} P_{in}^K(x) = \frac{1 - e^{-\frac{q}{p}\gamma\pi r^2}}{1 - e^{-\gamma\pi r^2}} = P_{in}; \tag{2.10}$$

$$\lim_{K \to \infty} P_c^K(x) = \frac{1 - e^{-(1-\rho)\gamma\pi r^2}}{1 - e^{-\gamma\pi r^2}} = P_c. \tag{2.11}$$

Proof. To obtain $P_{in}^K(x) \leq P_{in}$, $P_c^K(x) \leq P_c$, note that $f(x) = \ln(1 - \frac{q}{p}(1 - e^{-x})) + \frac{q}{p}x$ is an increasing function. In fact,

$$f'(x) = \frac{\frac{q}{p}(1 - \frac{q}{p})(1 - e^{-x})}{1 - \frac{q}{p}(1 - e^{-x})} \geq 0.$$

Then $f(x) \geq f(0) = 0$. Hence, we get that

$$\ln(1 - \frac{q}{p}(1 - e^{-\gamma s_i(x)})) \geq -\frac{q}{p}\gamma s_i(x).$$

So $\forall K \geq 1$,

$$\ln h^K \geq \sum_{i=1}^{K} -\frac{q}{p}\gamma s_i(x) = -\frac{q}{p}\gamma\pi r^2,$$

and it follows that $P_{in}^K(x) \leq P_{in}$. A similar derivation shows that $P_c^K(x) \leq P_c$.

---

[3]The proof goes through even if the clusters have significant different sizes.

We now show that $q/p \leq P_{in}^K(x)$, $1 - \rho \leq P_c^K(x)$. We state the result: For $a \geq 0, b \geq 0, a + b = 1, 0 \leq x_i \leq 1$, and for any positive integer K, the following inequality holds:

$$\prod_{i=1}^{K}(a + bx_i) \leq a + b\prod_{i=1}^{K} x_i. \tag{2.12}$$

We prove the above result by mathematical induction. When $K = 1$, the equality obviously holds. Assume that (2.12) holds for $K = n$. When $K = n + 1$, we have

$$\prod_{i=1}^{n+1}(a + bx_i) \leq (a + b\prod_{i=1}^{n} x_i)(a + bx_{n+1})$$

$$= a^2 + abx_{n+1} + ab\prod_{i=1}^{n} x_i + b^2\prod_{i=1}^{n+1} x_i$$

$$\leq a^2 + ab(1 + \prod_{i=1}^{n+1} x_i) + b^2\prod_{i=1}^{n+1} x_i$$

$$= a + b\prod_{i=1}^{n+1} x_i.$$

Then for $K = 1, 2, \cdots$, (2.12) holds. Using (2.12), we have that

$$\prod_{i=1}^{K}(\frac{s}{p} + \frac{q}{p}e^{-\gamma s_i(x)}) \leq \frac{s}{p} + \frac{q}{p}\prod_{i=1}^{K} e^{-\gamma s_i(x)} = \frac{s}{p} + \frac{q}{p}e^{-\gamma \pi r^2},$$

which shows that $q/p \leq P_{in}^K(x)$. A similar computation yields the result that $1 - \rho \leq P_c^K(x)$.

To prove Eqs. (2.10) and (2.11), let $h^K = \prod_{i=1}^{K}(1 - \frac{q}{p}(1 - e^{-\gamma s_i(x)}))$. By taking the log of h^K, we obtain $\ln h^K = \sum_{i=1}^{K} \ln(1 - \frac{q}{p}(1 - e^{-\gamma s_i(x)}))$. We have

$$\lim_{s_i(x) \to 0} \frac{\ln(1 - \frac{q}{p}(1 - e^{-\gamma s_i(x)}))}{-\frac{q}{p}\gamma s_i(x)}$$

$$= \lim_{s_i(x) \to 0} \frac{e^{-\gamma s_i(x)}}{1 - \frac{q}{p}(1 - e^{-\gamma s_i(x)})} = 1$$

where the first equation holds by l'Hôpital's rule. Then

$$\ln(1 - \frac{q}{p}(1 - e^{-\gamma s_i(x)})) = -\frac{q}{p}\gamma s_i(x) + o(s_i(x)).$$

When $K \to \infty$ and max $A_i \to 0$, then max $s_i \to 0$. Let C denote the circular region centered at x. We have

$$\lim_{K \to \infty} \ln h^K = \lim_{K \to \infty} \sum_{i=1}^{K} (-\frac{q}{p} \gamma s_i(x) + o(s_i(x))).$$

By the definition of integrals, we can get

$$\lim_{K \to \infty} \ln h^K = \iint_C -\frac{q}{p} \gamma ds = -\frac{q}{p} \gamma \pi r^2.$$

Let $\ln h = \lim_{K \to \infty} \ln h^K$, then $h = e^{-\frac{q}{p} \gamma \pi r^2}$. The result for (2.10) directly follows. A similar derivation yields the result for (2.11).

Remark. Other distributions of sensor placement, e.g., a uniform grid placement, could be used, and the results in Theorems 2.3–2.5 can be readily adapted based on the density of sensors in a unit area.

2.7   Coordinated Sleep Under Periodic Scheduling

In a dense network, there is significant spatial overlap in the sensing regions of sensors. Temporally, the on periods of the periodic sensors may also overlap. The latter problem is the most severe in the case of the synchronous network, but it is also a concern in the asynchronous network particularly when q/p is large. Such overlap leads to coverage redundancy and waste of energy. When a significant fraction of the sensors die, the coverage of the network degrades severely.

We propose a solution in which sensors exchange information about their locations, residual energies, etc., so that a sensor, say i, whose sensing region is completely covered by those of its active neighbors can go to sleep to conserve energy. Before i can go to sleep, it needs permission from its neighbors because they have to agree to remain active and cover i's responsibilities. Sensors renegotiate the permission to sleep from time to time. The decision to grant permission is partly based on the residual energies of the competing nodes, which allows sensors to rotate their roles and achieve an energy balance for maximum network lifetime.

We now present a *coordinated sleep protocol* (CSP) for sensors to achieve the goals above. The protocol can be applied to the synchronous, regionally synchronous, and asynchronous networks. CSP is similar to the RACP protocol in [3], but our main purpose is to study its use in a periodic scheduling context. In Sect. 2.8, we will evaluate how the two techniques interact in several design points. We will also present performance comparisons with RACP where periodic scheduling can achieve significant energy savings with competitive performance.

In CSP, an active sensor can assume one of three roles: *regular*, *supporting*, and *redundant*. Sensors periodically exchange hello messages indicating their ids, locations, period start times, roles, and residual energies.[4] The locations of the sensors can be obtained by GPS or a localization protocol [14, 15]. A supporting sensor has agreed to be active for a stated amount of time to support the sleep of a neighbor. While active in supporting role, the sensor maintains a timer that signals the end of this role. When the timer expires, the sensor changes role to regular.

For a regular sensor, say i, each subset of its active neighbors whose sensing regions completely cover that of i is called a *support* of i. Sensor i periodically checks if it has a non-empty support set. If so, i can choose a support set by (C1) the minimum residual energy of its members, denoted by E^{res}, or (C2) the overlap, denoted by L, between the intersection of the members' on periods and i's own on period. The first criterion aims to maximize the energy balance of the network, while the second criterion aims to maximize the productivity of the sleep being negotiated. There is generally a tradeoff between the two criteria, which can thus be integrated into a hybrid measure, denoted by ω, as a function of E^{res} and L, e.g., a weighted sum of the two measures. Ideally, sensor i should choose a minimal support set that scores highest in the chosen criterion (i.e., E^{res}, L, or ω). However, a large number of candidate support sets may exist. If sensor resources are limited, a more efficient algorithm that approximates the ideal selection can be used by i for the actual support set selection, as follows. (i) Via the hello messages, i learns from each neighbor j E_j^{res} (the remaining energy of j) and L_j (the overlap between i's and j's on periods). If needed, i can then compute ω_j as the function of E_j^{res} and L_j that integrates the criteria C1 and C2 above. (ii) i sorts the neighbors in decreasing order of their scores in the support set selection criterion. (iii) i includes into the support set the smallest number of the sorted list's top neighbors such that their combined sensing ranges subsume that of i.

Once i has chosen its support set, it broadcasts a request-to-sleep (RTS) packet to the neighbors in the support. Each neighbor who does not desire to go to sleep immediately replies to i with a clear-to-sleep (CTS) packet. Once i receives a CTS from all the support's members, it broadcasts a confirm (CNF) packet to the support, changes to redundant role, and sleeps for L time, after which it changes back to regular. While still waiting for at least one more CTS, i may receive an RTS from a neighbor, say j. In this case, i sets a (per-neighbor) random delay timer T_j for j. If, by the time T_j expires, i is still waiting for a CTS, i sends j a CTS. Once i receives a CNF from j, it assumes supporting role for j for the specified time. An illustration of the protocol is shown in Fig. 2.2.

Note that if L is less than $\frac{2c}{k_1-k_2}$, the sleep is too short to be productive for energy savings. In this case, i will simply decide not to send an RTS. This situation can be common if q is small (particularly for asynchronous networks), so that the temporal

---

[4]If the on periods of two sensors do not overlap, they do not hear each other's hello messages. This does not affect performance since there is no temporal overlap of coverage between the two sensors.

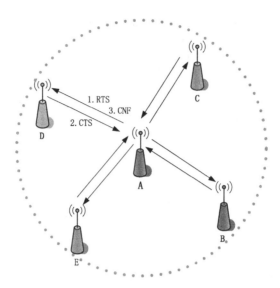

Fig. 2.2 Illustration of CSP protocol. The regular sensor A desires to go to sleep, and obtains permission by the following protocol message exchanges. (i) A sends a request-to-sleep (RTS) message to its supporting neighbors; (ii) The neighbors B–E send back a clear-to-sleep (CTS) to A if they can assume the supporting role; and (iii) A sends a confirm message (CNF) to B–E to indicate that it has obtained the required support

redundancy is not clear enough for coordinated sleep to help even if the sensor density is high. Another issue is i's setting of the random delay T_j. The design principle is that if i has a larger residual energy than that of j, then i should be more likely to send j the CTS before j sends i the CTS. Hence, it should be likely that i's delay timer for j is smaller than j's timer for i. We choose to pick T_j uniformly at random from $[0, 2^{E_j^{res}/E_i^{res}}]$, but alternative methods are possible.

Apart from support for periodic scheduling, CSP differs from RACP in its use of the delay timer. In RACP, a countdown delay is mandatory for every RTS. In CSP, i sends an RTS without delay, but a neighbor j who receives the RTS performs the random countdown if (and only if) j also desires to go to sleep. Hence, a countdown delay is avoided in CSP if there is no competition between neighbors in their sleep requests. Also, RACP uses a random sleep duration after a sensor has obtained permission to sleep, whereas CSP uses L as determined by the overlapping on periods of the sensors. Our experiments show that RACP's performance is more random in terms of variable network lifetimes achieved, while the lifetime of CSP is more predictable.

2.8 Numerical Results

We present Matlab results to illustrate and verify the analysis. In addition, we present Matlab simulations to evaluate the performance of the synchronous and asynchronous networks with and without coordinated sleep, including performance comparisons with RACP. Lastly, we will evaluate the regionally synchronous network under different sizes of the synchronization region, and compare its performance with the synchronous and asynchronous networks under coordinated sleep. We do not consider the energy costs of running either the CSP or RACP protocol. Hence, our comparison between CSP and RACP is fair. Further, the performance gains for both protocols are optimistic and can be regarded as upper bounds. Our results will show that in our synchronous, asynchronous, and regionally synchronous networks, the simplified best-case performance of CSP generally leads to fractional gains in system performance only, although these gains are indeed affected by different coordinated sleep opportunities in the different types of networks. Thus our simplification in the CSP evaluation will not affect the qualitative conclusion that the periodic scheduling, considered both locally for individual sensors and globally among the sensors, has a generally higher impact on system performance than CSP.

Unless otherwise stated, we use the following: (i) $X \in$ Exponential(α), $Y \in$ Exponential(β), $\alpha = 1$, and $\beta = 2$, where the distributions are numbers in time units of 0.1 h; (ii) for the energy model, $k_1 = 2.369$ J/h, $k_2 = 0.17$ J/h, and $c = 0.05$ J; (iii) each sensor of sensing range $r = 1$ m has an energy capacity of 9.26 mAh(equivalent to 100 J assuming a voltage of 3 V), and the sensors are deployed in a 20×20 m region according to a Poisson point process of intensity $\gamma = 4$, where γ is the average number of sensors per m^2; and (iv) for the distribution of PoIs, the deployment region is discretized into cells of dimensions 0.2×0.2 m and the center of each cell is a PoI. Each simulation in Sect. 2.8.1 is repeated until the standard deviation of the measurements is negligible compared with the average, and we report the average in the presentation. Representative traces are reported in Sect. 2.8.2.

2.8.1 Illustration of Analytical Results

2.8.1.1 Delay of Capture

Our approach allows a $P_c - P_{in}$ fraction of events to be captured with positive delay. Theorem 2.2 and Eqs. (2.2) and (2.3) quantify the expected delays D_{del} and D_{tot}. Figure 2.3a plots the analytical results for D_{del} against $p - q$ for different α. The measured averages in simulations are also shown as the indicated data points. Notice that the delay increases as the mean event staying time $\frac{1}{\alpha}$ increases. Also, when $p-q$

is large enough, most events that arrive early in an absent period do not stay long enough to be captured. Hence, the expected delay does not further increase with $p - q$ after some point.

Numerical results for the expected delay of all captured events, D_{tot}, are shown in Fig. 2.3b. Different from Fig. 2.3a, D_{tot} first increases and then decreases in Fig. 2.3b. To see how including D_{tot} in the network optimization may impact the results, we also plot the Q_E in Theorem 2.1 in Fig. 2.3b. Notice that for $p \in [0, p_1]$, both Q_E and D_{tot} increase. Hence, there is a tradeoff between Q_E and D_{tot} as the optimization goals. For $p \in (p_1, p_2]$, Q_E decreases while D_{tot} increases. Hence, this range of p is not optimal. For $p \in [p_2, \infty]$, both Q_E and D_{tot} decrease, again showing a tradeoff between the two optimization goals.

2.8.1.2 P_{in} of Asynchronous Network

Figure 2.3c verifies the correctness of Theorem 2.3 by plotting P_{in} against q/p for different γ. Notice the close agreement between the analytical and simulation results. More importantly, notice that at high sensor density (e.g., $\gamma = 4$), the achieved P_{in} is extremely close to 1 even if there is a significant fraction of off time in the periodic schedule (e.g., $\frac{q}{p} = 0.4$). *This shows that periodically resting the sensors does* not *necessarily cause noticeable delay in the event capture.*

2.8.1.3 Q_E of Synchronous Network

Figure 2.3d plots P_c (Theorem 2.1) against p for different values of q/p. Data points from simulations are also shown, which are in strong agreement with the analysis. For a fixed q/p, the function is monotonically decreasing in p, showing that we should turn on the sensor as briefly as possible at the start of each period. This is because once the sensor detects an event, the event is considered captured and the sensor does not gain by observing the event any longer. When energy is also considered, however, the Q_E plots given by (2.6) (Fig. 2.3e) initially increase with p and then decrease after reaching a global maximum. (The validation of Q_E follows directly from the validation of P_c.) Hence, the method outlined in Sect. 2.4 can be used to find the optimal p that maximizes Q_E.

2.8.1.4 Q_E of Asynchronous Network

We now illustrate Q_E, computed via the expression for P_c in Theorem 2.4, for the asynchronous network. We plot Q_E against p in Fig. 2.3f for different q/p. Notice that the plots are similar in trend to the Q_E plots of the synchronous network (Fig. 2.3e), although the peak performance is much less pronounced in this case. In particular, when $\frac{q}{p}$ is large (e.g., $\frac{q}{p} \geq \frac{1}{3}$), the Q_E value remains close to the optimal as p increases further.

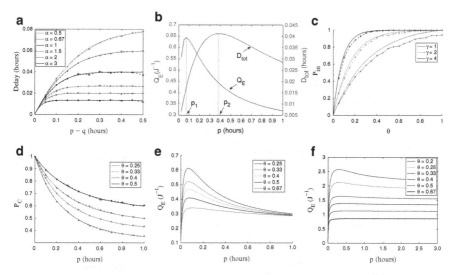

Fig. 2.3 Numerical results. (**a**) For events captured with positive delay, the average delay, D_{del}, against $p - q$ for varied α. (**b**) Q_E and D_{tot} against p for $\theta \triangleq \frac{q}{p} = 0.25$. (**c**) Plot of P_{in} of asynchronous network against $\theta \triangleq \frac{q}{p}$ for different γ. (**d**) Monotonically decreasing P_c of synchronous network against p for different $\theta \triangleq \frac{q}{p}$. (**e**) Plot of Q_E of synchronous network against p for different $\theta \triangleq \frac{q}{p}$. Optimal Q_E is achieved at an intermediate p. (**f**) Q_E of asynchronous network against p for different $\theta \triangleq \frac{q}{p}$. Peak of Q_E is less pronounced than synchronous network

2.8.2 Network Simulations

2.8.2.1 Asynchronous Network

We evaluate the asynchronous network with and without coordinated sleep by CSP. We assume that the user desires a stringent P_{in}^{min} close to 1, so that the required $\frac{q}{p} = 0.4$ by Theorem 2.3. Given $\frac{q}{p} = 0.4$, $p^* = 0.4$ h by optimizing Q_E numerically according to (2.6). Hence, $q^* = 0.16$ h.

As a representative trace, Fig. 2.4a shows the achieved P_{in} as a function of the deployment time for RACP and A-CSP. Notice that A-CSP has quite similar performance as RACP before either network dies, but A-CSP has about 74 % longer network lifetime. When the RACP network dies, the coverage drops quickly to zero. Death of the A-CSP network is much more gradual. After A-CSP starts dying, it takes about 30 h more before the network dies completely. Plots of P_c for the two networks are very similar to the corresponding P_{in} plots.

Figure 2.4b shows the achieved P_{in} for RACP and the asynchronous network *without* coordinated sleep (i.e., A-nc case). In this case, the asynchronous network has quite similar performance as RACP before either network dies, but it lasts about 74 % longer than RACP. Hence, A-nc lasts about as long as A-CSP before A-CSP

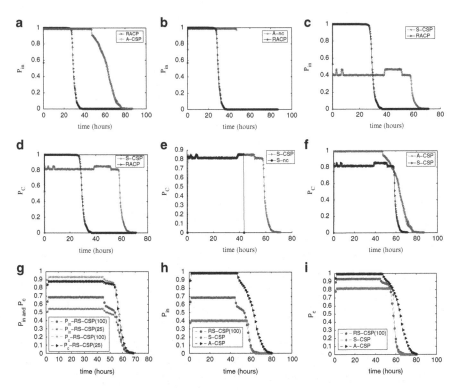

Fig. 2.4 Simulation results. (**a**) Achieved P_{in} against deployment time for RACP and A-CSP. A-CSP has longer lifetime and more gradual death than RACP. (**b**) P_{in} vs. deployment time for RACP and A-nc. A-nc runs longer than RACP. Its death is instantaneous since all the sensors work equally hard and run out of energy simultaneously. (**c**) P_{in} against deployment time for S-CSP and RACP. S-CSP achieves $P_{in}^{\min} = 0.4$ as specified by the user, and runs longer. (**d**) P_c vs. deployment time for S-CSP and RACP. S-CSP achieves $P_c = 0.8$ (cf. $P_{in} = 0.4$) and performs closer to RACP in terms of P_c. (**e**) P_c against deployment time for S-CSP and S-nc. CSP prolongs the network lifetime and achieves good energy balance between the sensors in the synchronous network. (**f**) P_c against deployment time for A-CSP and S-CSP. S-CSP operates longer before it starts to die and its death is less gradual than A-CSP. However, A-CSP performs better in terms of capture performance. (**g**) P_c and P_{in} as a function of the deployment time for the regionally synchronous network denoted as RS-CSP(K), where the number of clusters K is set to be 25 and 100 in two runs. (**h**) P_{in} against the deployment time for the synchronous, regionally synchronous, and asynchronous networks. (**i**) P_c against the deployment time for the synchronous, regionally synchronous, and asynchronous networks

starts to die. When A-nc dies, however, the coverage immediately drops to zero. This is because all the sensors work equally hard and run out of energy at the same time. Therefore, the usefulness of CSP in the asynchronous network lies mainly in its graceful degradation, where partial (but decreasing) coverage remains available over a significantly longer time duration. Because P_{in} is close to 1 for both networks, performance in terms of P_c is practically the same as P_{in}.

2.8.2.2 Synchronous Network

We now evaluate the synchronous network. We assume that the user is willing to relax the requirement for instantaneous event capture, and set $P_{in}^{\min} = 0.4$. Hence, $\frac{q}{p} = 0.4$. Optimizing p for Q_E in (2.6) yields $p^* = 0.075$ h. Hence, $q^* = 0.03$ h. Figure 2.4c compares RACP with S-CSP in terms of P_{in}. Notice that the network lifetime of S-CSP is about 2.1 times that of RACP. However, RACP has maximum instantaneous event-capture performance before it dies, whereas S-CSP achieves the relaxed P_{in} specified by the user (0.4). The performance comparison between RACP and S-CSP in terms of P_c is shown in Fig. 2.4d. Note that S-CSP has performance fluctuating between 0.81 and 0.85, which reflects the periodic on/off schedule of the network. Hence, the performance gap with RACP closes significantly in terms of P_c. This is because the synchronous network is designed to exploit the possibility of capturing an event before the event leaves a PoI. Hence, S-CSP allows the tradeoff of performance (17 % lower for P_c) for longer network lifetime (more than 2 times as long).

To evaluate the impact of coordinated sleep on the synchronous network, we compare the performance of S-CSP and S-nc in terms of P_c. The results are shown in Fig. 2.4e. Notice that (i) S-CSP lasts about 34 % longer than S-nc before S-CSP starts to die, and (ii) the death of S-nc is abrupt since the sensors work equally hard and they run out of energy at the same time, whereas the death of S-CSP is relatively more gradual.

2.8.2.3 Synchronous/Asynchronous Network Comparison

We now compare the synchronous and asynchronous networks under coordinated sleep. For A-CSP, we use the same evaluation case as in Sect. 2.8.2.1, i.e., $P_{in}^{\min}$ close to 1 and $\frac{q}{p} = 0.4$. For S-CSP, we either (i) keep $P_{in}^{\min}$ close to 1, in which case $\frac{q}{p}$ is also close to 1 (since $P_{in} = \frac{q}{p}$ in the synchronous network); or (ii) we keep $\frac{q}{p} = 0.4$, in which case we can only satisfy $P_{in}^{\min} = 0.4$. In case (i), the performance of S-CSP is similar to RACP's (although as Sect. 2.7 observes, the network lifetimes of RACP are more variable than S-CSP over different runs) since the sensors are active almost all the time. Hence, A-CSP performs better than S-CSP, as it performs better than RACP. In case (ii), it is clear that A-CSP performs better than S-CSP in terms of P_{in}. To see how A-CSP and S-CSP compare in terms of P_c and the network lifetime, refer to Fig. 2.4f. Note that S-CSP keeps the network running longer before it starts to die and the death is more abrupt, showing that S-CSP can achieve a better energy balance between the sensors. A-CSP starts dying sooner but takes a longer time to reach complete death. Considering the event capture performance throughout the deployment, A-CSP performs better than S-CSP overall.

2.8.2.4   Regionally Synchronous Network

We proceed to evaluate the performance of regionally synchronous networks. Since a main property of these networks is that they provide different levels of coordinated sleep opportunities, all the experiments in this section are conducted with CSP enabled.

We begin by evaluating the impact of the size of the synchronization region. We perform two experiments in which the network is divided into 25 and 100 clusters, respectively. For simplicity, we use a square network area, and divide it into regular 5×5 and 100×100 grids, respectively. We then use each cell in the grid to represent a cluster. The parameters p and q/p are set to be 0.075 and 0.4, respectively, which are the same as the corresponding S-CSP parameters used in the earlier experiments. The results comparing the numbers of clusters used are shown in Fig. 2.4g. It is clear that P_{in} and P_c are better in the 100-cluster case than in the 25-cluster case throughout the experiment. This shows that the better coordinated sleep opportunities in the 25-cluster network are not sufficient to offset the more effective event capture performance of the 100-cluster network (as discussed in Sect. 2.6). Hence, a larger number of clusters generally gives better overall performance when all the factors are considered. This general trend is true over a range of cluster sizes we have used in other experiments that are not reported in this chapter for simplicity.

We now compare the performance of the synchronous, regionally synchronous, and asynchronous networks. For the asynchronous network, we use the optimal parameters $p = 0.4$ and $q/p = 0.4$ as prescribed by the optimization procedure in Sect. 2.4. For the synchronous network, we use the optimal parameters $p = 0.075$ and $q/p = 0.4$. For the regionally synchronous network, we use the same setting as the synchronous network, i.e., $p = 0.075$ and $q/p = 0.4$. The P_c results for the three kinds of networks are shown in Fig. 2.4i, and the corresponding P_{in} results are shown in Fig. 2.4h. From the figures, we can see that the synchronous network starts dying later than the other two networks and achieves the best energy balance among them, showing that coordinated sleep is indeed more effective when the on periods of the sensors are more synchronized. However, similar to the previous experiments, this advantage is not enough to compensate for the synchronous network's reduced event capture performance as quantified by the previous analytical results. More generally, even when coordinated sleep is taken into account, the regionally synchronous network achieves an intermediate overall performance between the synchronous and asynchronous networks, and the asynchronous network remains the best option among the different kinds of networks.

2.8.3   Summary of Experiments

The experiments in Sect. 2.8.1 verify the analysis in Sects. 2.3–2.5 and illustrate the optimization of Q_E. The experiments in Sect. 2.8.2 compare S-nc, S-CSP,

A-nc, A-CSP, and RACP in a dense sensor network. We show that coordinated sleep in S-CSP can prolong the network lifetime compared with S-nc. S-CSP provides a performance/energy tradeoff versus RACP. It achieves the P_{in} specified by the user, and its achieved P_c can be significantly higher than P_{in} for events that stay. Moreover, the average delay of events captured can be quantified and used in the network optimization, although including the delay as an optimization objective may lead to tradeoffs with the Q_E objective, so that a user-specified balance between the two goals will be needed. We show that asynchronous periodic scheduling can achieve a higher global coverage intensity than S-CSP by spreading out the sensors' on periods. Importantly, the global operation can allow it to achieve maximum performance even in terms of *instantaneous* event capture (i.e., P_{in} close to 1) while being much more energy-efficient than RACP. A-CSP can further prolong the network lifetime over A-nc. It can achieve significantly longer network lifetimes than RACP with negligible loss of performance. More generally, A-CSP performs better than any regionally synchronized network, although this class of networks provides a broad spectrum of tradeoff between event capture performance and the effectiveness of coordinated sleep. Hence, A-CSP appears to be the best overall method in terms of maximum network lifetime and extremely good event capture performance.

2.9 Conclusions

We have investigated the use of periodic sensor scheduling for capturing stochastic events. We started off with the observation that when events can stay for some time, a (q, p)-periodic sensor can capture a fraction of events much higher than q/p. This is possible because events that arrive when a sensor is inactive may still be captured with a delay. We thus expect a tradeoff between energy efficiency (e.g., smaller q/p) and event capture (e.g., capture delay). We verify that such a meaningful tradeoff exists in the synchronous network relative to RACP. A similar tradeoff exists in the asynchronous network when the sensor density is moderately high, but the tradeoff becomes more attractive because the probability of non-instantaneous event capture decreases exponentially with $\frac{q}{p}$ in the asynchronous case (cf. linear decrease in the synchronous case). When the sensor density is high, the tradeoff becomes unnecessary in that the asynchronous network can achieve P_{in} close to one at high energy efficiency. When P_{in} is close to one, the performance is in fact not dependent on the event dynamics. We also considered a class of regionally synchronous networks where the size of the synchronization region is specifiable. This class of networks allows to control the synchronization overhead, and represents a broad spectrum of tradeoff between the effectiveness of event capture and that of coordinated sleep. We have evaluated the effectiveness of CSP for overall energy savings in the synchronous, regionally synchronous, and asynchronous networks.

References

1. S. He, J. Chen, D. Yau, H. Shao, and Y. Sun. Energy-efficient capture of stochastic events under periodic network coverage and coordinated sleep. *IEEE Transactions on Parallel and Distributed Systems*, 23(6):1090–1102, 2012.
2. I. Dietrich and F. Dressler. On the lifetime of wireless sensors networks. *IEEE Transactions Sensor Networks*, 5(1):1–38, Jan. 2009.
3. C. Hsin and M. Liu. Network coverage using low duty-cycled sensors: random and coordinated sleep algorithms. In *Proceedings of ACM/IEEE International Conference on Information Processing in Sensor Networks (IPSN)*, 2004.
4. N. Bisnik, A. Abouzeid, and V. Isler. Stochastic event capture using mobile sensors subject to a quality metric. In *Proceedings of the Annual International Conference on Mobile Computing and Networking (MobiCom)*, 2006.
5. D. Yau, N. Yip, C. Ma, N. Rao, and M. Shankar. Quality of monitoring of stochastic events by periodic and proportional-share scheduling of sensor coverage. In *Proceedings of The International Conference on emerging Networking EXperiments and Technologies (CoNEXT)*, 2008.
6. N. Jaggi, K. Kar, and A. Krishnamurthy. Rechargeable sensor activation under temporally correlated events. *Springer Wireless Networks*, 15(5):619–635, 2009.
7. S. Chellappan, X. Bai, B. Ma, D. Xuan, and C. Xu. Mobility limited flip-based sensor network deployment. *IEEE Transactions on Parallel and Distributed Systems*, 18(2):199–211, 2007.
8. X. Li, H. Frey, N. Santoro, and I. Stojmenovic. Strictly localized sensor self-deployment for optimal focused coverage. *IEEE Transactions on Mobile Computing*, 10(11):1520–1533, 2011.
9. T. He, C. Huang, B. Blum, J. Stankovic, and T. Abdelzaher. Range-free localization and its impact on large scale sensor networks. *ACM Transactions on Embedded Computing Systems*, 4(4):877–906, 2005.
10. Crossbow, http://www.xbow.com. *Crossbow MPR/MIB Users' Manual*.
11. K. Shenai and S. Mukhopadhyay. Cognitive sensor networks. In *Proceedings of International Conference on Microelectronics (MIEL)*, 2008.
12. M. Franceschetti, O. Dousse, D. Tse, and P. Thiran. Closing the gap in the capacity of wireless networks via percolation theory. *IEEE Transactions on Information Theory*, 53(3):1009–1018, 2007.
13. S. Ganeriwal, R. Kumar, and M. B. Srivastava. Timing-sync protocol for sensor networks. In *Proceedings of the International Conference on Embedded Networked Sensor Systems (SenSys)*, 2003.
14. D. Evans and L. Hu. Localization for mobile sensor networks. In *Proceedings of the Annual International Conference on Mobile Computing and Networking (MobiCom)*, 2004.
15. K. Yedavalli and B. Krishnamachari. Sequence-based localization in wireless sensor networks. *IEEE Transactions Mobile Computing*, 7(1):1–14, Jan. 2008.

Chapter 3
Energy-Efficient Trap Coverage in Sensor Networks

3.1 Introduction

While recent advances in wireless communication and hardware device have posed a bright blueprint for Wireless Sensor Network (WSN) applications in a large range of fields including military affairs, health care and environment surveillance [1–3], most practical implementations are restricted to small-scale experiments or applications with dozens or hundreds of sensors. One of the major reason for such relatively small scale deployment is the prohibitively high cost of deploying thousands/millions of sensor nodes for large scale applications. Consequently, designing a sensor network where the number of sensor nodes does not increase quickly (e.g., exponentially) with the deployment size while maintaining desired system performance is a fundamental challenge.

Partial area coverage is introduced in [4] to address the limited quantity of sensor nodes in large-scale applications, as it is prohibitively expensive to guarantee the full coverage of the Region of Interest (RoI) [5–8]. *Partial area coverage* allows coverage holes [9] and its quality of coverage is mainly indicated by the ratio of uncovered area to the whole region [4, 10, 11]. By adopting partial area coverage scheme, a significant number of sensor nodes could be saved and thus scale well with the network size. However, for the partial area coverage, it is difficult to evaluate the actual network performance by the ratio of uncovered area as the size of coverage holes could be extremely large or even unbounded. For applications such as intrusion detection, this implies that the moving intruder (target) could travel an arbitrarily long distance and time in certain area of the RoI without being detected. Therefore, the simple partial area coverage has a very narrow spectrum of practical applications that has been well recognized by many researchers.

To trade off the scalability and network performance, P. Balister et al. recently propose a new kind of area coverage, called *trap coverage* [12], based on the concept of *partial area coverage*. In *trap coverage*, the size of each coverage hole is incorporated as the indicator of quality of sensing. For the coverage hole, its size is indicated by its *diameter D*, which is defined as the largest Euclidean distance

S. He et al., *Energy-Efficient Area Coverage for Intruder Detection in Sensor Networks*,
SpringerBriefs in Computer Science, DOI 10.1007/978-3-319-04648-8_3,
© The Author(s) 2014

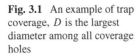

Fig. 3.1 An example of trap
coverage, D is the largest
diameter among all coverage
holes

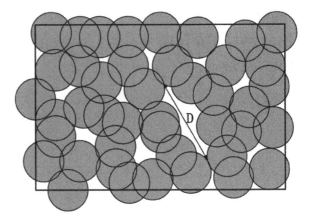

between any two points in the coverage hole. A set of sensor nodes is said to
provide *D-Trap Coverage* to the RoI A, if the diameter of every coverage hole
in A is smaller or equal to D. Although there may exist lots of coverage holes in
the network according to this definition, the largest area of coverage holes is no
more than πD^2. An example of *trap coverage* is illustrated in Fig. 3.1. A target is
trapped in a coverage hole as it will be detected within a maximal moving distance
of D. Compared with the partial coverage which takes coverage ratio as an indicator,
trap coverage guarantees the quality of coverage in the worst case. By carefully
controlling the parameter D, network performance such as connectivity or delay of
detecting intrusions can be ensured in *trap coverage*.

P. Balister et al. for the first time study the trap coverage of randomly deployed
sensor networks. They consider the fundamental problem of how to design reliable
and explicit deployment density required to achieve *D-trap coverage*. Their work
is concerned with conceptual network design, however, practical implementation
scheme such as how to simultaneously guarantee trap coverage and energy efficiency is left uninvestigated. As sensor nodes could be deployed in an arbitrary
manner, the required number of sensor nodes to ensure *D-trap coverage* is usually
more than the optimal value.

In this chapter, we consider the energy-efficient scheduling of sensor nodes in
the randomly deployed sensor networks to achieve *D-trap coverage*. In fact, this
problem is extremely difficult and we have proved that it is an NP-hard problem.
To effectively solve this problem, we design an approximation algorithm called *Trap
Cover Optimization* and implement the algorithm in a localized way. Specifically,
the main intellectual contributions of this chapter are as follows [13]:

1. We study how to schedule the activation of sensors to maximize network lifetime
   while guaranteeing *D-trap coverage* in the randomly deployed sensor networks.
   Efficient algorithms have been designed to solve the problem in polynomial time.
2. We first introduce a centralized solution and theoretically prove that the performance of this centralized algorithm is not greater than $O(\rho)$ times of the optimal
   solution, where ρ is the density of sensors scattered in the RoI, i.e., ρ is defined

as the ratio of the number of sensor nodes N to the size of RoI S. Our algorithm attains a provable guarantee in the worst case which is only related to ρ. The approximation ratio will be more desirable when our algorithm is applied to large-scale wireless sensor networks.

3. To make our design practical for real sensor network systems, we introduce a localized trap coverage protocol that can be implemented at individual sensor nodes.

4. Extensive simulations are performed to demonstrate that our proposed algorithm is effective and much more energy efficient than a naive approach to the optimal lifetime scheduling for trap coverage problem, as well as the state-of-the-art solutions.

The rest of the chapter is organized as follows. We formulate the *Minimum Weight Trap Cover Problem* in Sect. 3.2. Section 3.3 presents the details of our algorithm design. The approximation ratio of the algorithm is obtained theoretically in Sect. 3.4. We design a localized protocol in Sect. 3.5, perform extensive simulations to verify the effectiveness of algorithm in Sect. 3.6 and conclude the chapter in Sect. 3.7.

3.2 Preliminary and Problem Formulation

3.2.1 Network Model

We consider a large scale WSN in the RoI A. We assume A is a rectangle of size $S = l_1 \times l_2$ for simplicity. The WSN consists of N sensor nodes. Assume that the location of each sensor is known, either by GPS or other localization techniques [14, 15]. The location of individual sensor nodes is denoted by P_i, where $i = 1, 2, \cdots, N$. Each sensor can only detect targets in a certain range, which we refer to as sensing region, denoted by R. Each sensor is assigned with a unique ID number. As in the previous chapter, we assume the sensing regions of sensor nodes are homogeneous, and all are unit *open* disc centered at the location of the sensor with radius of r, which does not include the boundary of sensing region [16, 17]. By making such simplified assumption, we can concentrate on our main problem and understand its intrinsical property, with a minor penalty of performance degradation in practical applications. The boundary of sensing region R_i of sensor i is referred to as *sensing border*, which is essentially a circle of radius r centered at P_i.

For large scale sensor network applications, controlled deployments of sensor nodes are normally infeasible, therefore lead to the popular adoption of random deployment. For example, an airplane can be used to airdrop sensor nodes in a forest. So for our network, we assume sensor nodes are randomly deployed with a density of ρ. Apparently, ρ can be approximated by N/S, where $S = l_1 \times l_2$ is the area of region A. While a classic Poisson point process is assumed in [12], our solution can be applied to all potential deployment processes.

We divide operation time of individual nodes into time slots. At each slot, a subset of sensors is activated to ensure trap coverage. We rotate active time of sensor nodes in different slots in order to extend network lifetime. Assume that each sensor has an initial energy of E units, $E > 1$, and consumes 1 unit per slot if it is active. For simplicity, if a sensor is put in *sleep* mode, we assume it consumes no energy. The sensor node with residual energy less than 1 unit can not be activated any more.

In terms of communication, each sensor node can only communicate with other sensor nodes within a certain range, referred as *transmission range*. As proved in [18], if the transmission range of sensor node is at least twice of its sensing range, coverage implies connectivity of the network. This is to say if the sensing region of two sensors intersects with each other, they are connected. In trap coverage, the sensing regions of isolated sensors do not intersect with each other, meaning the isolated set of sensors must be trapped in a coverage hole with uncovered physical points around. If we neglect the detection measurements of isolated sensors, the main connected component of sensors can still provide required *D-trap coverage* (see the definition in the Sect. 3.2.2). We therefore assume that trap coverage also implies connectivity of the network.

3.2.2 Trap Coverage Model

Trap coverage is a new coverage model allowing the existence of uncovered physical points in the RoI but restricts the size of coverage holes, as shown in Fig. 3.1. In this section, we give a mathematical definition of *trap coverage*.

Definition 3.1 (Coverage hole). A set of uncovered points in RoI A forms a coverage hole H, if for any two points a and b in H, there always exists a curve ζ whose start and end points are a and b, respectively, satisfying that $\zeta \cap (\bigcup_{i \in C} R_i) = \emptyset$, where C is the set of active sensor nodes at that time. Obviously, $H \subseteq A$.

The diameter of coverage hole H is defined as the largest Euclidean distance between any two points in the coverage hole. Denote the diameter of coverage hole H by $d(H)$ and $dist(a,b)$ is the Euclidean distance between point a and b, then

$$d(H) = \max_{a,b \in H} (dist(a,b)). \tag{3.1}$$

Definition 3.2 (D-trap coverage). Sensor set C provides *D-trap coverage* to RoI A, if the diameter of every coverage hole H in A is not greater than D, i.e.,

$$d(H) \le D, \forall H \subseteq A. \tag{3.2}$$

We call C a *D-trap cover* of RoI A.

Obviously, if we set diameter threshold D to zero, *D-trap coverage* reverts back to full coverage.

3.2.3 Minimum Weight Trap Cover Problem

To achieve energy efficiency, we aim to design sensing scheduling algorithms that activate the minimal number of sensor nodes per time slot while guaranteeing D-trap coverage. To guarantee the balance of energy consumption among all sensors, we also require that the activated sensors should be those with more residual energy.

Each sensor in the network is assigned with a weight based on its residual energy at the beginning of each time slot. We show an example of weight assigning in this chapter. Let E_i denote the residual energy of sensor node i and $\gamma_i = 1 - E_i/E$ denote its energy consumption ratio, where, γ_i is a variable between 0 and 1. The weight of sensor node i at time slot $t, t = 1, 2, \cdots$, is assigned as an exponential function related to the residual energy, i.e.,

$$w_i(t) = \theta^{\gamma_i(t)}/E, 0 \le \gamma_i < 1 \tag{3.3}$$

where θ is a constant value greater than one. Sensor i is specially marked by assigning weight θN which is greater than the sum of weights of other sensors with residual energy, if it has no residual energy, i.e., $\gamma_i = 1$. The weight θN can also be viewed as an infinite high weight. After weight assigning, we formulate *Minimum Weight Trap Cover Problem* in this section.

Consider a sensor set C which provides D-*trap coverage* to RoI A, i.e., each sensor i in C is associated with a weight w_i. The weight of trap cover C is defined as the sum of weights of all sensors in C, i.e., $w(C) = \sum_{i \in C} w_i$. Given the diameter threshold D, there exists a family of trap covers $\aleph$.

Definition 3.3 (Minimum Weight Trap Cover). Given RoI A, a set of sensors $\{1, 2, \cdots, N\}$ with corresponding weights $w_1, \cdots, w_N$. A minimum weight trap cover C^* is a *trap cover* with minimum weight among all trap covers, i.e.,

$$w(C^*) = \min_{C \in \aleph} w(C) = \min_{C \in \aleph} \sum_{i \in C} w_i. \tag{3.4}$$

Minimum Weight Trap Cover Problem: Given RoI A, a set of sensors $\{1, 2, \cdots, N\}$ with their corresponding weights $w_1, w_2, \cdots, w_N$ and sensing radius r. C is a subset of sensors. There is M coverage holes $H_1, \ldots, H_M$ in RoI A if all sensors in C are activated while other sensors not in C are put into sleep. The *minimum weight trap cover problem* is to choose a minimum weight set C^* which can ensure that every coverage hole in A has a diameter no more than D, where D is a threshold set by applications.

The problem can be formally formulated as follows.

$$\begin{aligned} \min_{C \in \aleph} \quad & \sum_{i \in C} w_i \\ s.t. \quad & d(H_m) \le D, m = 1, \cdots, M. \end{aligned} \tag{3.5}$$

3.3   Algorithm Design

In this section, we show the concept of intersection point and then discuss how to calculate the diameters of coverage holes in the RoI. Based on the results, we further design our algorithm, *Trap Cover Optimization*, for minimum weight trap cover problem.

3.3.1   Finding the Diameter of a Coverage Hole

We will show how to calculate the diameters of coverage holes in this section to prepare for the design of trap coverage optimization algorithm.

An *intersection point* is one of the two points where two sensors' sensing boundaries intersect with each other. We would like to introduce some basic knowledge on intersection points in previous literatures before presenting an efficient solution to the minimum weight trap cover problem. Let Ω denote the set of intersection points of all sensors' sensing boundaries in RoI.

Firstly, a sensor set covers every point in region A if and only if it covers all points in set Ω.

The theorem above is first presented and proved in [19]. Considering the sensing region of a sensor is a disc, the problem of region coverage can be easily transformed into the problem of finding a vertex cover [20]. Note that the sensing region of a sensor is *open* which does not include the boundary of sensing region, the intersection points marked by red dot in Fig. 3.2 are still not covered.

Recall that the diameter of a coverage hole is defined as the largest Euclidean distance between any two points in the coverage hole. Let Ω_a denote the set of intersection points of all active sensors' sensing boundaries. Without loss of generality, we also consider the situation on the edge of A by adding intersection points among all active sensors' sensing boundaries and the boundary of A into set Ω_a. The set Ω_{H_m} denotes the intersection points on the border of H_m.

Secondly, the diameter of coverage hole H_m equals to $d(\Omega_{H_m})$ if the sensing regions are convex, i.e.,

$$d(H_m) = d(\Omega_{H_m}) = \max_{a,b \in \Omega_{H_m}} dist(a,b), \tag{3.6}$$

where $d(\Omega_{H_m})$ denotes the largest Euclidean distance between any two points in set Ω_{H_m}.

The theorem above and detailed proof are first presented in [12]. We can therefore calculate the diameter of a coverage hole in a convenient way. An example is demonstrated in Fig. 3.2. In Fig. 3.2, the boundary of coverage hole H is composed of sensors' sensing boundaries and $\Omega_{H_m} = \{P_1, \cdots, P_8\}$, are the intersection points of these sensing borders. The diameter of coverage hole H equals to the maximum Euclidean distance of the intersection points in Ω_{H_m}, i.e., $d(H) = \max_{a,b \in \Omega_{H_m}} dist(a,b)$.

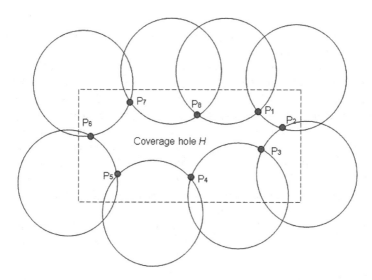

Fig. 3.2 A demonstration of diameter calculation. The diameter of coverage hole H equals to the maximum Euclidean distance between the intersection points a and b in point set Ω_{H_m}, i.e., $d(H) = \max_{a,b \in \Omega_{H_m}} dist(a, b)$, hereby $\Omega_{H_m} = \{P_1, P_2, \cdots, P_8\}$ is the set of intersection points on the boundary hole H

3.3.2 Algorithm Overview

The threshold D in Minimum Weight Trap Cover Problem, which indicates the quality of coverage, is determined by the applications, so we need to guarantee the threshold D in the algorithm design. At first, we present the following theorem.

Theorem 3.1. *Minimum Weight Trap Cover Problem is NP-hard.*

Proof. Assume the minimum distance between any two intersection points in RoI is ϵ. Assume the parameter D in Minimum Weight Trap Cover Problem which satisfies $\epsilon > D > 0$. If there exists a coverage hole with diameter M in the RoI, then $M \geq \epsilon > D$ since M is the distance between two intersection points and it should be no less than ϵ. Thus, there exists no coverage hole in the RoI if D-trap coverage is guaranteed. We need to cover all intersection points with active sensors to provide D-trap coverage in this case.

In this special case, i.e., $\epsilon > D > 0$, Minimum Weight Trap Cover Problem is reduced to a set covering problem which has already been proved to be NP-hard [21]. As a conclusion, Minimum Weight Trap Cover Problem (for any $D > 0$), as a more general problem, should be no less hard than the set covering problem. So we can claim that Minimum Weight Trap Cover Problem is NP-hard.

Since the problem is NP-hard and can not be precisely solved in polynomial time unless $P = NP$, we develop an efficient approximation algorithm *Trap Cover Optimization* (TCO) to solve minimum weight trap cover problem. Assume C' is the *minimum weight sensor cover* [22] which is a set of sensors providing full coverage to A, and C is the trap cover which provides D-trap coverage to A. We will derive C from set C' in TCO, i.e., $C \subseteq C'$.

TCO is composed of two major steps. Firstly, a *minimum weight sensor cover* C' is selected to cover the whole RoI A, which is viewed as Minimum Weight Sensor Cover Problem [22]. Given RoI A, a set of sensors $s_1 \dots s_n$ and monitored subregion and the weight w_i for each sensor, the problem is to find a set of sensors with minimum total weight. Let Ω denote the set of intersection points of all sensors' sensing boundaries in RoI A. We regard all intersection points in Ω as targets to be covered. Existing literatures have developed algorithms to efficiently solve this problem [5, 19, 20, 23]. We will introduce DSC proposed in [19] and use it to find a minimum weight sensor cover in Step 1 in Sec. 3.5.1. The output of TCO, C is empty initially. We let Ψ represent the $C \cup C'$ to simplify our description.

Secondly, we remove each sensor in C' successively and consider whether to put it into C. Consider a sensor i in C'. Given the sensors in Ψ are active, sensor i is added into C when the maximum coverage hole diameter will exceed the threshold D if i is inactive.

Finally, C' is empty and C, the output of TCO, equals Ψ. TCO activates the sensors in C. As we can see in the procedure of TCO, given that the sensors in Ψ are active, sensor i is removed from Ψ only when the maximum coverage hole diameter will not exceed the threshold D if i is set to be inactive. Our aim is to remove as much sensors with poor residual energy as possible in Ψ and activate only a few sensors which are rich in residual energy. The challenge is to design an optimal removal strategy and remove sensors in a proper order so that we can achieve a better performance.

3.3.3 Removal Strategy Design

We will discuss how to design the removal strategy in this section. Again, We let Ψ represent the $C \cup C'$ to simplify our description. To remove more sensors from Ψ, we consider the possible impact of removing a candidate sensor on the diameters of existing coverage holes. The diameters of coverage holes will increase or remain unchanged when we remove a sensor from Ψ. Intuitively, if we always remove sensors which will cause the diameters of existing coverage holes increase quickly, the largest diameter will soon approach the threshold D, thus only a few sensors can be removed. More sensors can be removed if we choose to remove sensors with less impact on the diameters of existing coverage holes. To quantify and bound the impact, we introduce $T_\Psi(i)$ and $D_\Psi(i)$ as the upper bounds of increase on the diameter of coverage hole when removing a sensor node i. At first, let $\Omega_\Psi(i)$ represent all intersection points which are covered by set Ψ but not covered by set $(\Psi - i)$. Assume points in set $\Omega_\Psi(i)$ belong to boundary points of M_i coverage holes. Accordingly, we divide $\Omega_\Psi(i)$ into $\Omega_{\Psi_1}(i), \Omega_{\Psi_2}(i), \cdots, \Omega_{\Psi_{M_i}}(i)$. Assume the diameters of these coverage holes are $d_{\Psi_1}(i), \cdots, d_{\Psi_{M_i}}(i)$, respectively.

$T_\Psi(i)$ denotes the aggregate diameters of all coverage holes which are covered by sensor i but not covered by set $(\Psi - i)$, i.e., the sum of diameters of all newly emerging coverage hole when sensor i is removed. We set

$$T_\Psi(i) = \sum_{j=1}^{M_i} d_{\psi_j}(i) \tag{3.7}$$

Actually $T_\Psi(i)$ is the largest possible increment of a coverage hole when i is removed from Ψ. We illustrate that $T_\Psi(i)$ is an upper bound of increase on the diameter of coverage hole in Fig. 3.3. As shown in Fig. 3.3, there exists a coverage hole with diameter d_1. The dots denote the intersection points. The sensor i denoted by dashed circle is about to be removed and $T_\Psi(i)$ is marked in the figure. Assume the diameter of coverage hole after removing i is d_2. We introduce an auxiliary line d_q which connects two intersection points in the initial coverage hole in Fig. 3.3. d_1 is greater than d_q due to the definition of diameter of coverage hole. $d_2 < d_q + T_\Psi(i)$ holds because of the triangle inequality. Thus, we have $d_2 - d_1 < T_\Psi(i)$, which suggests that $T_\Psi(i)$ is an upper bound of increment of a coverage hole when i is removed.

Note that $T_\Psi(i)$ should not be the maximum diameter among all newly emerging holes because these holes may merge into one large coverage hole when removing other sensors. Actually, the impacts should be bounded by summing up the diameters.

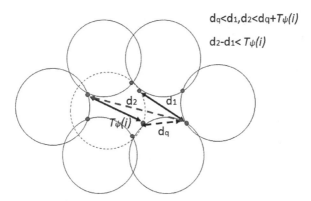

$d_q < d_1, d_2 < d_q + T_\Psi(i)$

$d_2 - d_1 < T_\Psi(i)$

Fig. 3.3 $T_\Psi(i)$ is an upper bound of increment of a coverage hole when i is removed. There exists a coverage hole with diameter d_1. The *dots* denote the intersection points. The sensor i denoted by *dashed circle* is about to be removed. Assume the diameter of coverage hole after removing i is d_2 and d_q is an auxiliary line

We set $D_\Psi(i)$ as Eq. (3.8).

$$D_\Psi(i) = \min\{T_\Psi(i), 2r\} \tag{3.8}$$

As shown in Fig. 3.3, $T_\Psi(i)$ is an upper bound of increment of a coverage hole. Meanwhile, the increment of a coverage hole when removing a sensor should not be greater than the diameter of sensing region. Thus, we have Eq. (3.8) as a more strict upper bound for the increase on diameter of coverage hole, which is important to determine how to remove sensors with less impact on the diameter of coverage holes.

$D_\Psi(i)$ is an important metric because it denotes the possible impact of removing a candidate sensor on the diameters of existing coverage holes. Since $D_\Psi(i)$ represents the increment of diameters of coverage holes when sensor i is removed from C, we can preferably remove sensors with low $D_\Psi(i)$ whose effects on the existing coverage holes are bounded. Consequently, potentially more sensor nodes can be removed before the diameter of any coverage hole is beyond D. We adopt $D_\Psi(i)$ in TCO as an important factor. We also conduct simulation experiments to justify our design. Sensors with same weights are deployed randomly and a full cover set C' is picked with aforementioned methods. We compare the amount of sensors removed from Ψ between TCO and a random approach which randomly selects a sensor to remove. The average results are plotted in Fig. 3.4, which shows that TCO improves the amount of removed sensors significantly by considering $D_\Psi(i)$.

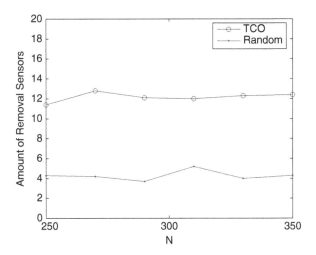

Fig. 3.4 Amount of removed sensors by TCO and random approach vs. N, $D = 25$

Besides the amount of sensors in C, we also need to minimize the weights of sensors in C, so we consider to normalize the weights of sensors by $D_\Psi(i)$ to determine which sensor is to be removed. $D_\Psi(i)$ is a variable between 0 and $2r$. To avoid zero in denominator, we set the normalized factor as $1/(1 + \alpha D_\Psi(i))$, where $\alpha = 1/(2r)$. Furthermore, the normalized weight $G(i)$ of sensor i is defined, i.e.,

$$G(i) = w_i/(1 + \alpha D_\Psi(i)). \tag{3.9}$$

We remove the sensor with the greatest normalized weight $G(i)$ each time. If there exist sensors with same $G(i)$, we remove the sensor with the lowest ID number. In this way, sensors with less residual energy or less upper bound of increment of a coverage hole, i.e., $D_\Psi(i)$, are supposed to be removed from Ψ with a higher priority. Every sensor i in set C' are checked iteratively and added to C if the maximum coverage hole diameter of $(\Psi - i)$ exceeds the threshold D, while the uncovered intersection points and coverage holes are updated accordingly. TCO terminates when C' is empty. The remaining set C is the output of TCO.

Sensors with no residual energy are not involved in minimum weight sensor cover C' since they have infinite weights, unless there are no set covers with residual energy to provide full coverage. Even if sensors with no residual energy are involved in C', they are removed first in TCO because they have infinite normalized weights. If the output of TCO, C contains sensors with no residual energy, it indicates that there are no trap covers with redundant energy to provide D-trap coverage any more, which means that the network reaches the end of its lifetime.

Remark. Though we assume that the sensing radius of sensors are equal as a common scenario for simplicity, TCO can also be applied if the sensing ranges of sensors are unequal. Assume r_{max} as the maximum sensing radius among all sensors in this case. We only need to replace the radius r by r_{max} in TCO to guarantee the upper bound. That is to say, Eq. (3.8) should be $D_\Psi(i) = \min\{T_\Psi(i), 2r_{max}\}$ and α in Eq. (3.9) should be $1/(2r_{max})$. The rest of TCO does not need to be modified for an unequal sensing radius case.

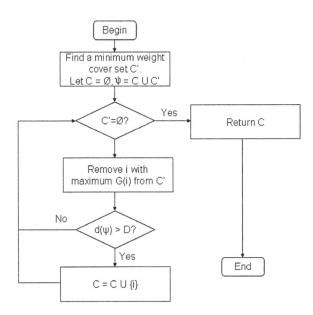

Fig. 3.5 An illustration of Algorithm 1

3.3.4 Algorithm Illustration

We illustrate TCO in Fig. 3.5 to help understand the algorithm. The detailed algorithm is shown in Algorithm 1 and $d(\Psi)$ is used to represent the maximum diameter of coverage holes when only sensors in set Ψ are activated.

A simple example of TCO is presented in Fig. 3.6. The four sensors in set C' are deployed symmetrically which fully cover the square region with side length a and set C is empty initially. The sensors are assumed to have the same weights. The threshold of coverage hole is supposed to be less than a. We will show how TCO works then. At first, we assume TCO picks sensor 1 to be removed from set C'. Since $d(\Psi)$ where $\Psi = C' \cup C$ is not beyond the threshold, sensor 1 will not be added into set C. Next, we find that sensor 3 has the lowest $D_\Psi(i)$ among C' and Ψ only contains sensor 2, 3, 4 now. Considering the sensors have the same weights, we will remove sensor 3 from C'. After that, $d(\Psi)$ is still not beyond the threshold, so sensor 3 will not be added into set C either. We then remove sensor 2 from set C'. The diameter of Ψ which only contains sensor 4 can not provide required trap coverage any more, so we add sensor 2 into set C. In the same way, we remove

---

Algorithm 1 Trap cover optimization

1. Find a *minimum weight cover set* C' which ensures the whole region A is covered. Let $C = \emptyset, \alpha = 1/(2r)$. Let Ψ represent $C' \cup C$.
2. For every sensor i in set C', calculate $G(i) = w_i/(1 + \alpha D_\Psi(i))$. If $C' = \emptyset$, return trap cover C.
3. Find the sensor i with maximum $G(i)$ and remove i from C'.
4. Update existing uncovered intersection points in A and the boundaries of coverage holes with respect to set Ψ.
5. Calculate $d(\Psi)$. If $d(\Psi) > D$, then let $C = C \cup \{i\}$.
6. Back to step 2.

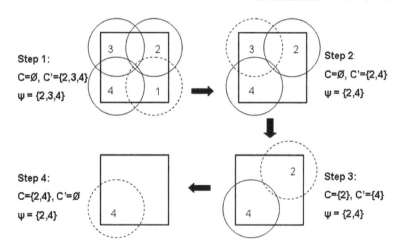

Fig. 3.6 An example of Algorithm 1

sensor 4 from C' and add it into set C. Finally, TCO terminates when C' is empty and sensor 2 and 4 in C are activated to provide required trap coverage.

The time complexity of TCO is apparently polynomial since we only traverse the elements in C' once. Later on, we will prove the approximation ratio of TCO is only related to the sensor deployment density in Sect. 3.4. Simulations in Sect. 3.6 have confirmed that $G(i)$ based TCO always picks trap cover with higher average residual energy and lower energy consumption.

3.4 Performance Analysis

3.4.1 Theoretical Analysis

We investigate the performance of our proposed algorithm TCO theoretically in this section.

Before the derivation, we make assumptions as follows.

Assumption 3.1. *Given RoI A of size $l_1 \times l_2$, $l_1 \gg r + D$ and $l_2 \gg r + D$, where r is the sensing range of each sensor and D is the diameter threshold of D-trap coverage.*

We will first prove the ratio bound of aggregate weight between the output C and the initial input C' of TCO. Let $N_{C'}$ denote the number of sensors in C'. w_C and $w_{C'}$ denote the aggregate weight of sensors in C and C' respectively.

Lemma 3.1. $w_C \leq \frac{2N_{C'}}{2N_{C'} + D/(2r)} w_{C'}$.

Proof. Let $\overline{C} = C' - C$ denote the set of sensors which are removed from set C' by Algorithm 1. Here we use $D(i)$ to represent $D_\Psi(i)$ for simplicity. Suppose at the $(k+1)_{th}$ iteration $d(C'_{k+1})$ exceeds threshold D for the first time. Let Q denote the set $C' - C'_k$, where C'_k denotes C' at the k_{th} iteration. Thus, $C \subseteq C'_k$. Obviously, $Q \subseteq \overline{C}$, which means $w_Q \leq w_{\overline{C}}$.

Since TCO always selects to remove sensor i from C' with maximum $G(i) = w_i/(1 + \alpha D(i))$, we get that,

$$\frac{w_Q}{\sum_{i \in Q}(1+\alpha D(i))} \geq \max_{j \in C}\{\frac{w_j}{1+\alpha D(j)}\} \\ \geq \frac{w_C}{\sum_{j \in C}(1+\alpha D(j))} \tag{3.10}$$

According to the definition of set Q and $D(i)$, the upper bound of the incremental of maximum coverage hole diameter, we have

$$\sum_{i \in Q} D(i) \geq D - 2r \tag{3.11}$$

With Eqs. (3.10), (3.11) and $\alpha = 1/(2r)$,

$$
\begin{aligned}
\frac{w_{\overline{C}}}{w_C} &\geq \frac{w_Q}{w_C} \\
&\geq \frac{\sum_{i \in Q}(1 + \alpha D(i))}{\sum_{j \in C}(1 + \alpha D(j))} \\
&\geq \frac{\alpha D}{(1 + 2\alpha r)N_{C'}}
\end{aligned}
\tag{3.12}
$$

With the Eq. $w_{C'} = w_{\overline{C}} + w_C$, we have

$$
\frac{w_C}{w_{C'}} \leq \frac{(1 + 2\alpha r)N_{C'}}{(1 + 2\alpha r)N_{C'} + \alpha D}
\tag{3.13}
$$

$$
w_C \leq \frac{2N_{C'}}{2N_{C'} + D/(2r)} w_{C'}
\tag{3.14}
$$

where concludes the proof.

We denote the optimum minimum weight trap cover which provides D-trap coverage as OPT. Let N_1 denote the number of sensors in OPT.

Lemma 3.2. *The number of sensors providing D-trap coverage to RoI A of size $l_1 \times l_2$ must be greater than $\frac{2S}{3\sqrt{3}(r+D)^2}$ where $S = l_1 \times l_2$.*

Proof. Suppose that there exists a point P in a coverage hole. The distance between P and the nearest boundary of detection area is less than D. Assume the boundary is belonged to sensor i. Then the distance between P and sensor i is less than $(r + D)$ according to the triangle inequality and P will be covered if the sensing distance of i increases to $(r + D)$. For all sensors, if the sensing radius r is increased to $(r + D)$, the sensor set will provide full coverage to A.

Fig. 3.7 Optimal deployment on the vertices of equilateral triangles

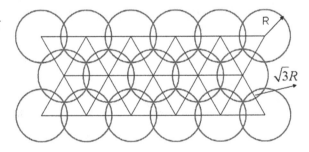

It is well-known that it is optimal to deploy sensor nodes of disk sensing model on the vertices of equilateral triangles to cover a plane [24] (see an illustration in Fig. 3.7). If $l_1 \gg r + D$ and $l_2 \gg r + D$, according to the property of equilateral triangles, the minimum number of sensors with sensing range $r + D$ which provide full coverage to the RoI A is $\frac{2S}{3\sqrt{3}(r+D)^2}$. We have

$$N_1 \geq \frac{2S}{3\sqrt{3}(r+D)^2}. \tag{3.15}$$

This concludes the proof.

Assume that w_{OPT} denotes the aggregate weight of sensors in set OPT. According to Eq. (3.3), the weight of energy-redundant sensor i, $w(i)$ satisfies that $\theta/E > w(i) \geq 1/E$. We have

$$w_{OPT} \geq N_1/E. \tag{3.16}$$

$$\frac{w_{C'}}{w_{OPT}} < \frac{N\theta/E}{N_1/E}$$
$$\leq \theta N \frac{3\sqrt{3}(r+D)^2}{2S}. \tag{3.17}$$
$$= \rho\theta \frac{3\sqrt{3}(r+D)^2}{2}$$

We have the following main result for TCO, which theoretically guarantees the performance of TCO even in the worst case. Based on Lemma 3.1 and Eq. (3.17), we have the following theorem.

Theorem 3.2. $w_C/w_{OPT} < \frac{2N_{C'}}{2N_{C'}+D/(2r)}\rho\theta\Phi$, where $\Phi = \frac{3\sqrt{3}(r+D)^2}{2}$.

As θ, r and D are constants, the approximation ratio of TCO is only related to the density ρ. As the number of sensors in a full cover set $N_{C'}$ increases, the approximation ratio approaches $\rho\theta\Phi$, which is treated as $O(\rho)$. The bound guarantees the approximation ratio of TCO compared with optimal solution even in the worst case. A better ratio is possible for special cases, but in the worst case, our approximation ratio holds. If more and more sensor nodes are placed, the optimal solution improves quickly since more options are available. Compared with optimal solution, our algorithm relatively improves slower. Thus, the worst bound may deteriorate as ρ increases. But the performance of our algorithm is still desirable, since the density of nodes will not be extremely high for real deployment.

We also plot a figure to show the magnitude of the ratio intuitively. If we set the average amount of sensors to cover the region $N_{C'}$ is 20, $D = 5, r = 15, \theta = 1.01$, we have the approximation ratio curve in Fig. 3.8. For the sensing radius r which is 15, ρ above 0.01 is relatively high in reality since there are almost 7 sensors within the sensing range of a sensor whose area is πr^2.

Remark. As we remark in Sect. 3.3.3, TCO can be applied even if the sensing ranges of sensors are unequal. Assume that r_{max} is the maximum sensing radius among the sensing ranges of all sensors. In this case, the approximation ratio of TCO is $\frac{2N_{C'}}{2N_{C'}+D/(2r_{max})}\rho\theta\Phi$, where $\Phi = \frac{3\sqrt{3}(r_{max}+D)^2}{2}$. The induction is very similar to the proof of Theorem 3.2, so we do not repeat here.

Fig. 3.8 Approximation ratio
vs. ρ, $N_{C'} = 20$, $D = 5$,
$r = 15$, $\theta = 1.01$

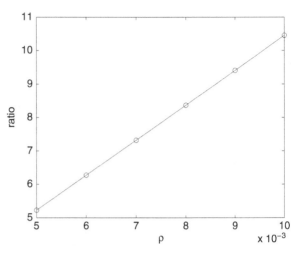

3.4.2 Network Lifetime Analysis

Given a minimum weight trap cover with an approximation ratio $O(\rho)$ acquired by
TCO, we can induct the approximation ratio bound of network lifetime if TCO is
performed every time slot. We refer to the framework of proof of network lifetime
approximation ratio in [19] to induct our approximation ratio of network lifetime
below.

Assume L is the network lifetime by TCO and L^* is the theoretical maximum
lifetime by optimal algorithm. If TCO activates sensors with residual energy, the
network is said to be *alive*; else if TCO attempts to activate sensors with no residual
energy, the network is *dead*. We define $\varsigma = \{1, \cdots, L\}$ as the set of time slots
when the lifetime of network is alive. Since L is no greater than L^*, we define
$\varsigma^* = \{L+1, \ldots, L^*\}$ as the set of slots when the network is dead under TCO
algorithm but might still stay alive under optimal algorithm.

Theorem 3.3. L^* *is at most an* $O(\rho)$ *factor greater than* L.

Proof. Assume trap cover set $C(t)$ is selected by TCO at each slot t and $C^*(t)$ is
the trap cover which is set to be active by optimal maximum lifetime algorithm.
$w_C(t)$ is the sum of the weights of sensors in set $C(t)$ at slot t and $w_C^*(t)$ is the sum
of the weights of sensors in set $C^*(t)$ during the period of running TCO each time
slot, $w_C(t) = \sum_{i \in C(t)} w_i(t)$ and $w_C^*(t) = \sum_{i \in C^*(t)} w_i(t)$.

As proved in Theorem 3.2, we have

$$w_C(t) \leq \rho\theta\Phi w_{C_{OPT}}(t) \leq \rho\theta\Phi w_C^*(t).$$

If the network is not alive, $t \in \varsigma^*$, we have $w_C(t) \geq \theta N$. Thus,

$$w_C^*(t) \geq \frac{N}{\rho\Phi} \quad \forall t \in \varsigma^*.$$

Define the function $X\{i \in C^*(t)\}$ as follows,

$$X\{i \in C^*(t)\} = \begin{cases} 1, \text{ if } i \in C^*(t) \\ 0, \text{ otherwise} \end{cases}$$

We can derive that

$$
\begin{aligned}
\tfrac{N}{\rho\Phi}(L^* - L) &\leq \textstyle\sum_{t \in \varsigma^*} w_C^*(t) \\
&= \textstyle\sum_{t \in \varsigma^*} \sum_{i \in C^*(t)} w_i(L+1) \\
&= \textstyle\sum_{t \in \varsigma^*} \sum_{i=1}^N w_i(L+1)X\{i \in C^*(t)\} . \\
&= \textstyle\sum_{i=1}^N w_i(L+1) \sum_{t \in \varsigma^*} X\{i \in C^*(t)\} \\
&\leq E \textstyle\sum_{i=1}^N w_i(L+1)
\end{aligned}
\tag{3.18}
$$

Because the network is dead when $t \in \varsigma^*$, any sensor $i, i \in C^*(t)$ does not consume energy, which means $w_i(L+1) = w_i(j), t \in \varsigma^*$. Inequality $\sum_{t \in \varsigma^*} X\{i \in C^*(t)\} \leq E$ holds, because active sensor i under theoretical optimal lifetime algorithm must not cost more energy than E.

At any slot $t \in \varsigma$, sensor $i \notin C(t)$ will not consume any energy, thus $w_i(t) = w_i(t+1)$. The aggregate weight $C(t) \sum_{i \in C(t)} \frac{\theta^{l_i(t)}}{E} \leq N\theta, t \in \varsigma$. Hence we get

$$
\begin{aligned}
&E \textstyle\sum_{i=1}^N (w_i(t+1) - w_i(t)) \\
&= E \textstyle\sum_{i \in C(t)} (w_i(t+1) - w_i(t)) \\
&= \textstyle\sum_{i \in C(t)} \theta^{l_i(t)} (2^{\frac{\log_2 \theta}{E}} - 1) \\
&\leq \log_2 \theta \textstyle\sum_{i \in C(t)} \frac{\theta^{l_i(t)}}{E} \\
&\leq N\theta \log_2 \theta
\end{aligned}
$$

We can know that

$$
\begin{aligned}
&E \textstyle\sum_{i=1}^N w_i(L+1) \\
&= E \textstyle\sum_{t=1}^L \sum_{i=1}^N (w_i(t+1) - w_i(t)) \\
&\leq \theta \log_2 \theta L N + E \textstyle\sum_{i=1}^N w_i(1) \\
&= N(L\theta \log_2 \theta + 1)
\end{aligned}
\tag{3.19}
$$

By combining Inequalities (3.18) and (3.19), we get

$$L^* \leq L(\rho\theta\Phi \log_2 \theta + 1) + \rho\Phi, \tag{3.20}$$

where Φ and θ can be seen as constants. The proof completes.

3.4.3 Simulation Performance

TCO removes as many sensors with high weight as possible. We conduct simulations to validate the performance of TCO, i.e., how much aggregate weight can be removed. The ratio of the aggregate weight of removed sensors to the aggregate weight of initial full cover sensor set is viewed as the indicator of the performance of TCO. We present the boxplot in Fig. 3.9a, which shows the statistics of running for 300 times to test the performance of TCO in average. As we can see, TCO performs well both in the average case and in the worst case. The removed aggregate weight ratio is even above 0.45 in the worst case, which guarantees the effect of employing TCO.

We also record the status of maximum hole diameter and aggregate weight of sensors in Ψ during a period of TCO running in Fig. 3.9b. The results illustrate the running status of TCO. The maximum hole diameter increases very slowly when it approaches D. That is because we remove sensors not just depend on $G(i)$. We enumerate each candidate sensor i successively according to the magnitude of $G(i)$ when the maximum hole diameter is about to exceed D. Candidate sensor i is removed only if it will not cause violation against restraint of D. Hence, many sensors are removed with no significant effect on maximum hole diameter at the end.

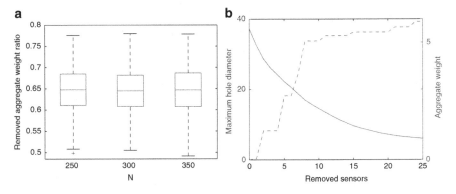

Fig. 3.9 The performance of TCO, D=40. (**a**) Removed aggregate weight ratio vs. N. (**b**) Status of maximum hole diameter and aggregate weight

3.5 Localized Protocol

3.5.1 Protocol

We have proposed and analyzed our algorithm TCO in Sects. 3.3 and 3.4. It is significant to implement our algorithm in a real WSN. Due to the nature of WSN, we design a localized protocol, *Localized Trap Coverage Protocol*, to implement TCO in WSN, which avoids global communication and thus improves energy efficiency of network.

Each sensor has two modes: sleep and active. At the beginning of each time slot, every sensor wakes up and performs distributed algorithm DSC in [19] to determine whether to be active initially. DSC is a distributed protocol that guarantees full coverage. Each sensor in DSC calculates the so-called *activation preference ratio* based on local information and determines whether to sleep according to the activation preference ratio. The upper bound of approximation ratio of the distributed algorithm is $O(\log n)$, the same as the well known centralized algorithm on set cover problem [25]. The active sensors then perform *Localized Trap Coverage Protocol* to find a minimum D-trap cover. We focus on the design of Localized Trap Coverage Protocol next.

Each sensor communicates with other sensors within a distance of $(r + D)$ in a multi-hop way where r is the sensing range and D is the diameter threshold. The sensors within a distance of $(r + D)$ from sensor i is defined as neighbor sensors of i, denoted by set C_i. Suppose that set V_i contains active sensors in C_i. Every active sensor broadcasts *Initial-Message* including location information at the beginning to recognize its neighbor sensors. Since the location of each sensor is known by itself, either by GPS or other localization techniques [14, 15], it is not difficult for sensor i to find its active neighbor sensors.

In TCO, the sensor with the greatest $G(i)$ in Eq. (3.9) has the highest priority to be removed, or to say, switch to sleep. If there exist sensors with same $G(i)$, we remove the sensor with the lowest ID number. Since the protocol is designed as localized implementation of TCO, we also introduce priority to schedule the sensors. Define $pr_i = \{G(i), ID_i\}$ as the priority of sensor i where ID_i is a unique number of sensor i and $G(i)$ is calculated by sensor i as Eq. (3.9). $pr_i > pr_j$ if, (i) $G(i) > G(j)$ or (ii) $G(i) = G(j)$ and $ID_i < ID_j$. Sensor i packs the priority pr_i into *Initial-Message*, and shares the information with its neighbor sensors. Thus, every sensor knows the initial priority of its active neighbor sensors after initialization.

Sensor i will send *Mode-Message* to broadcast its decision after it has chosen to sleep or stay active. There exist two kinds of Mode-Message: Mode-Message$_{sleep}$ and Mode-Message$_{active}$. If i receives Mode-Message$_{sleep}(j)$ from sensor j in V_i, it removes j from V_i. Suppose set M_i contains sensors in V_i with higher priority than sensor i which have not decided to sleep or stay active. If i receives any kind of Mode-Message from sensor j in M_i, it removes sensor j from M_i.

Sensor i starts to make decision when M_i is empty. If there does not exist a coverage hole whose diameter is greater than D in the area around sensor i within the range of $(r + D)$ which is covered by set V_i, it means the set V_i can guarantee D-trap coverage within the area. Then sensor i chooses to sleep to save energy. Note that sensor i is not in set V_i. If V_i is not sufficient to cover the area with no hole greater than the threshold, sensor i has to stay active since D-trap coverage can not be guaranteed without sensor i. Sensor i broadcasts Mode-Message and stays active or turns into sleep accordingly after decision making.

However, for sensor i, $G(i)$ maybe varies when a sensor whose sensing region overlaps with that of i chooses to sleep (see the definition of $G(i)$ in Eq. (3.9)).

When $G(i)$ varies, sensor i needs to broadcast the updated priority to its neighbor sensors in *Update-Message*, which is used to convey updated priority of sensors.

There is a useful property of the priority of sensors. Since $D_\Psi(i)$ will never decrease and w_i remains unchanged during a time slot, $G(i)$ can only decrease if it varies. Thus the priority of sensor i will not increase during a time slot. Based on the property of priority, sensor i only needs to consider Update-Messages from sensors in M_i since updated priority is always lower than original value. If sensor i receives Update-Message from sensor j in set M_i, it compares the updated priority with pr_i to decide whether to remove j from M_i. The set M_i may also be updated due to the change of pr_i to guarantee that M_i only contains sensors with higher priority than pr_i.

To help understand the procedure of the protocol, we illustrate the process of a normal sensor i in Fig. 3.10 for Localized Trap Coverage Protocol. At the first state, sensor i receives *Initial-Messages* from its neighbor sensors. Sensor i also broadcasts its own *Initial-Message* and other received messages whose senders are within a certain range. After a certain period of time, referred to as t_w, sensor i turns into the second state, which suggests that the initialization process ends. At the second state, sensor i receives *Mode-Messages* and *Update-Messages* from other sensors, and broadcasts its own *Update-Message* and other received messages whose senders are within a certain range. When i has the highest priority among all sensors within a certain range, it turns into the third state. At the third state, i decides whether to stay active or sleep and broadcast its decision in the *Mode-Message*. Then i turns into the fourth state, at which it acts as the decision, either stays active or switches to sleep.

The details are summarized in Algorithm 2. To avoid dead lock caused by message dropping in the network, we also set a timeout threshold t_τ which is

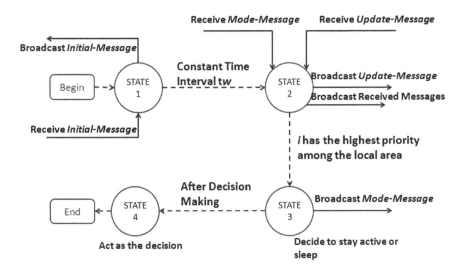

Fig. 3.10 The procedure of sensor i in *localized trap coverage protocol*

Algorithm 2 Localized trap coverage protocol

1: Define $pr_i = \{G(i), ID_i\}$ as the priority of sensor i. $pr_i > pr_j$ if $G(i) > G(j)$ or $(G(i) == G(j)$ and $ID_i < ID_j)$;
2: Set a timeout threshold t_τ.

   % *Find the set of active neighbor sensors, V_i*
3: Let $V_i = \emptyset$;
4: Broadcast *Initial-Message(i)*;
5: Set a time window with length t_w.
6: **while** Time window not end **do**
7:     **if** Receive new *Initial-Message(j)* and $distance(i, j) < r + D$ **then**
8:         Add j into set V_i;
9:         Broadcast *Initial-Message(j)*;
10:     **end if**
11: **end while**
   % *Wait for Mode-Messages and decide whether to stay active or sleep*
12: Define set M_i contains sensors in V_i whose priority is greater than pr_i;
13: **while** true **do**
14:     % *Update V_i and M_i when receiving Mode-Message from sensors in V_i*
15:     **if** Receive new *Mode-Message(j)* **then**
16:         **if** $distance(i, j) < r + D$ **then**
17:             Broadcast *Mode-Message(j)*;
18:             **if** j decides to sleep **then**
19:                 Remove j from V_i;
20:                 Calculate $G(i)$ again. Broadcast *Update-Message(i, pr_i)* and update M_i if $G(i)$ changes;
21:             **end if**
22:             **if** $j \in M_i$ **then**
23:                 Remove j from M_i;
24:             **end if**
25:         **end if**
26:     **end if**
27:     % *Update M_i when receiving Update-Message from sensors in M_i*
28:     **if** Receive new *Update-Message(j, pr_j)* **then**
29:         **if** $distance(i, j) < r + D$ **then**
30:             Broadcast *Update-Message(j, pr_j)*;
31:         **end if**
32:         **if** $j \in M_i$ and $pr_j < pr_i$ **then**
33:             Remove j from M_i;
34:         **end if**
35:     **end if**
36:     % *Clear M_i if timeout*
37:     **if** $M_i \neq \emptyset$ and exceed timeout threshold t_τ **then**
38:         $V_i = V_i - M_i$;
39:         Let $M_i = \emptyset$;
40:     **end if**
41:     % *Start to decide if M_i is empty*
42:     **if** $M_i == \emptyset$ **then**
43:         Given sensors in V_i are active;
44:         **if** There exists a coverage hole greater than D within the range of $(r + D)$ **then**
45:             Broadcast *Mode-Message_{active}(i)*;
46:             Stay Active Mode;
47:             break;
48:         **else**
49:             Broadcast *Mode-Message_{sleep}(i)*;
50:             Turn into Sleep Mode;
51:             break;
52:         **end if**
53:     **end if**
54: **end while**

---

determined by applications. In Algorithm 2, line 3–10 find the set of active neighbor sensors, V_i. For lines 11–53, sensor i waits for Mode-Messages and decides whether to stay active or sleep. In detail, lines 14–25 update V_i and M_i when receiving Mode-Message from sensors in V_i, lines 27–34 update M_i when receiving Update-Message from sensors in M_i, lines 36–39 clear M_i when the timeout threshold is exceeded, and finally lines 41–52 decide whether to sleep or stay active when M_i is empty.

The protocol schedules the activation of sensors to provide D-trap coverage at each time slot. All sensors make decisions based on the localized protocol at the beginning of a time slot and act accordingly during the time slot. We further provide theoretical analysis of the protocol in Sect. 3.5.2.

3.5.2 Analysis

We propose localized trap coverage protocol to implement our algorithm in WSN. The protocol is theoretically analyzed in this section. We prove that the protocol exactly implements TCO in a localized way. Besides, we also analyze the communication cost of the protocol.

Define the area around a sensor i within the range of $(r + D)$ as its neighbor region. The active neighbor sensors contained in V_i are all located in the neighbor region of i.

Assume the initial active sensor set which provides full coverage to the whole RoI is C'. The active sensor set acquired by localized trap coverage protocol is C_1 and sensor set activated by TCO is C_2. Apparently, $C_1 \subseteq C'$ and $C_2 \subseteq C'$. TCO removes sensors one by one, while localized trap coverage protocol is performed in a parallel way. We have Theorem 3.4 that the output of localized trap coverage protocol is the same as TCO, i.e., $C_1 = C_2$. To prove the theorem, we propose Lemma 3.3 at first.

Lemma 3.3. *If sensor i has the highest priority among its neighbor sensors, its decision will not be influenced by other sensors.*

Proof. For sensor i with the highest priority among its active neighbor sensors, i.e., $pr_i > max_{s \in V_i} pr_s$, the sensors in V_i will wait because they have a neighbor sensor i with a higher priority than themselves. Thus, the priority of sensor i will not change since no neighbor sensor changes the mode at this moment. As the priority will never increase, the sensors in V_i will always have a lower priority than i though the mode of their neighbor sensors may vary. Thus, the decision of sensor i will not be interfered by other sensors in V_i.

The active sensors which are not in V_i may change its mode at this moment. The decision of sensor i will not be influenced since it is based on whether the active sensors in V_i can provide D-trap coverage to the neighbor region or not, which has no relationship with sensors which are not in V_i.

We can conclude that the decision of sensor i is independent.

Lemma 3.3 confirms that it makes no difference to remove sensors with the highest priority among its neighbor sensors in a parallel way because removing each sensor is independent and there is no interference between them.

Theorem 3.4. *Localized trap coverage protocol and TCO provide the same output.*

Proof. Localized trap coverage protocol and TCO both put sensors into sleep in the order of priority which is calculated in the same way. A sensor with the highest priority in the whole region which is to be removed in TCO is also necessary to be put into sleep in localized trap coverage protocol because it must have the highest priority among its neighbor sensors. Assume the sequence of sensors in C' considered by TCO is $s_1, s_2, \cdots, s_k$ and their decisions are $D_1, D_2, \cdots, D_k$ respectively, where $D_i \in \{0, 1\}$. Sleep mode is denoted by 0 and active mode is denoted by 1. We show that even though localized trap coverage protocol may not put sensors into sleep in the same order as $s_1, s_2, \cdots, s_k$, the decisions of the sensors are the same as $D_1, D_2, \cdots, D_k$ by mathematical induction.

In localized trap coverage protocol, the decision of sensor s_1 is the same as D_1 because it has the highest priority. Assume that for an integer i, the decisions of $s_1, s_2, \cdots, s_i$ are the same as that of TCO, i.e., $D_1, D_2, \cdots, D_i$ respectively. We now prove that the decision of sensor s_{i+1} is D_{i+1} in localized trap coverage protocol.

Consider sensor $s_x, x > i + 1$, which has made decision before sensor s_{i+1}. s_x is not a neighbor sensor of s_{i+1}, otherwise s_x should had a higher priority than s_{i+1}, which violates the relationship in the sequence of TCO. Since the decision of s_{i+1} is only related with its neighbor sensors, s_x will not influence the decision of s_{i+1} according to Lemma 3.3. Thus, we only need to consider sensors which rank higher than s_{i+1} in the sequence of TCO, i.e. $\{s_1, s_2, \cdots, s_i\}$. Assume $Q = \{s_1, s_2, \cdots, s_i\}$ and the active neighbor sensors of s_{i+1} is $V_{s_{i+1}}$. The sensor set which influences the decision of s_{i+1} is $Q \cap V_{s_{i+1}}$. Sensors in $Q \cap V_{s_{i+1}}$ have the same decisions as theirs in TCO as we assumed. In this scenario, we assume $D_{i+1} = 0$ without loss of generality. The whole RoI should not emerge a coverage hole with diameter greater than D because of the removal of s_{i+1}, otherwise TCO should activate sensor $i + 1$ rather than $D_{i+1} = 0$. Sensor s_{i+1} will choose to sleep because no coverage hole greater than D exists in its neighbor region. If $D_{i+1} = 1$, there should exist a coverage hole with diameter greater than D when sensor s_{i+1} is removed. The part of the coverage hole in the neighbor region of sensor s_{i+1} should have a diameter greater than D, otherwise the mode of sensor s_{i+1} will not influence the diameter of the coverage hole. Thus, sensor s_{i+1} will stay active because a coverage hole with diameter greater than D will emerge within its neighbor region if it chooses to sleep. So we have conclusion that the decision of s_{i+1} is D_{i+1} in localized trap coverage protocol.

We can conclude that the decisions of all sensors are the same in both TCO and localized trap coverage protocol, which suggests that they have the same output.

Even though every sensor only guarantees that there is no coverage hole greater than D locally in the protocol, we have the Corollary 3.1 that the protocol leads to global D-trap coverage. It holds since localized trap coverage protocol activates the same sensors as TCO which guarantees global D-trap coverage.

Corollary 3.1. *Global D-trap coverage is guaranteed in Localized Trap Coverage Protocol.*

After proving localized trap coverage protocol is a localized implementation of TCO theoretically, we analyze the communication cost of the protocol to evaluate its performance as a localized protocol. Communication collision and message dropping are not considered in the analysis.

The set of initial active sensors can provide full coverage to the RoI. Suppose there exist ρ' active sensors in a unit area in the average case, i.e., the density of initial active sensors is ρ'. The communication cost of each active sensor i is represented by the amount of messages i that needs to broadcast. In the protocol, there exist three kinds of messages: Initial-Message, Update-Message and Mode-Message. For an individual active sensor i, it has one Initial-Message, one Mode-Message and several Update-Messages about itself to broadcast. Update-Message is only broadcasted when an active sensor whose sensing region overlaps with that of i switches to sleep, so sensor i has $4\pi r^2 \rho'$ Update-Messages in the average case. Thus, there are $(2 + 4\pi r^2 \rho')$ messages of i to broadcast in total during a time slot in the average case. Sensor i needs to broadcast each new message about sensor j it receives if j is within the neighbor region of i for the sake of multi-hop communication. On average, there exist $\pi(r + D)^2 \rho'$ active sensors within the neighbor region, so the amount of messages each sensor needs to broadcast is $(2 + 4\pi r^2 \rho')\pi(r + D)^2 \rho'$. ρ' is always constant since the set of initial active sensors is a minimum weight cover set which covers the RoI with minimum sensors. Thus, the communication cost mainly depends on D if r is constant. It increases quadratically as the increasing of D. With the formula, we can estimate the amount of transmission messages when we get a minimum weight sensor cover.

3.5.3 How to Find the Largest Diameter

In Algorithm 2, each sensor needs to calculate the diameter of every coverage hole within its neighbor region and check whether the diameter has exceeded the threshold. Though we can calculate the diameter of a coverage hole by the intersection points if its boundary is known, the challenge is that it is not easy to find all coverage holes and their boundaries given the deployment of activated sensors.

One way is to divide the region into grids and check whether each grid is covered or not. The uncovered grids merge into a coverage hole if they are physically next to each other. Thus, we can find all coverage holes and their boundaries. But we do not adopt the approach in our algorithm because it is only an approximation way and the computation cost is high. Actually, we employ a data structure called *disjoint-*

set to manage the coverage holes. In Algorithm 2, sensor i needs to check the largest diameter every time when a sensor in V_i chooses to sleep. Sensor i gradually updates the coverage holes with *disjoint-set* when removing a sensor in V_i instead of calculating all coverage holes with grid approximation each time.

In computing, a disjoint-set structure is a data structure that keeps track of a set of elements partitioned into a number of disjoint (non-overlapping) subsets [26]. It supports multiple operations like *union* and *add*. In the protocol, the elements in the set are uncovered intersection points and each subset is a coverage hole denoted by a set of intersection points on its boundary. The coverage holes are disjoint subsets. If any of them have common intersection points, which suggests that they are physically connected, they will merge into one coverage hole eventually. At the beginning of each time slot, there is no coverage hole and uncovered intersection point since we find a set of sensors providing full coverage. There is no subset and element at the beginning. When sensor i removes a sensor in V_i, there may emerge newly uncovered intersection points and coverage hole. If there emerge uncovered intersection points, we *add* the uncovered intersection points as elements and coverage holes as subsets into the disjoint set. After performing *add* action, we perform *union* action to merge coverage holes into a big one if any of them have common intersection points. By *add* and *union* operations in the disjoint set, we can denote each coverage hole in the region as different subsets of intersection points. If we want to calculate the diameter of a coverage hole, we just need to find the maximum distance between any two intersection points in the subset.

We show how to perform *add* and *union* operations here. Suppose we have several sensors in Fig. 3.11. The subset which already exists in the disjoint set is S_1 with uncovered intersection points $\{H_1, H_2, H_3, H_4\}$. If the sensor in dash line is supposed to be removed, there emerge a coverage hole and uncovered intersection points $\{H_3, H_4, H_5, H_6, H_7, H_8\}$ on its boundary. This new coverage hole should be merged into the existing coverage hole S_1 since they are physically connected. In our algorithm, we will add a new subset/coverage hole S_2 with intersection points $\{H_3, H_4, H_5, H_6, H_7, H_8\}$. Since S_1 and S_2 have joint elements H_3 and H_4, we

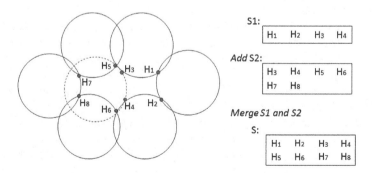

Fig. 3.11 Add and union operations in disjoint-set

merge these two sets into S as the merged coverage hole, and we can calculate the diameter of the merged coverage hole by finding the maximum distance between any two intersection points in the set S.

3.6 Simulation Results

3.6.1 Experiment Setup

The WSN in our simulations has N sensors, each with an initial energy of E units and a sensing range of 15 units. The sensors are deployed randomly in a square of 100×100 units^2. Active sensors in each time slot consume 1 unit of energy. We assume that the switching frequency is very low so that the communication costs in scheduling are negligible in the simulations. Otherwise, the energy of network would be wasted in the exchange of scheduling messages and switch of sensor mode between active and sleep. The diameter threshold of trap coverage is D units. We conduct simulations on Matlab 2010b.

The simulations are conducted mainly in following procedures. Firstly, at the beginning of each time slot, each sensor is assigned with a weight according to its residual energy. The sensor with more residual energy is assigned lower weight. Secondly, we employ specified algorithm to find a D-trap cover and only sensors in the trap cover are activated during each time slot. Finally, the lifetime of network terminates if there exists no trap cover with redundant energy to provide D-trap coverage any more.

3.6.2 Energy Balance and Consumption

In this section, we conduct extensive simulations to evaluate the performance of TCO in a lifetime span of WSN.

We choose two of the most efficient algorithms to the maximum network lifetime under full coverage model for comparison in the simulation part, *Greedy-MSC Heuristic* [5] and *DSC* [19]. These two approaches are both designed for the full coverage model. We also describe a naive approach designed for trap coverage requirement used in the simulation to compare with TCO. We name the naive approach as *Naive-Trap*, which is slightly modified from the Greedy-MSC Heuristic algorithm with adjustment to the trap coverage requirement. Assume $U(k) = z(k)/w_k$ for a sensor k where $z(k)$ is the number of uncovered intersection points covered by k and w_k is the weight of sensor k. This algorithm always sets the sensor k^* with maximum $U(k^*)$ to be active iteratively until the region achieves *D-trap coverage*. At each time slot, sensors are assigned with weights and a minimum weight trap cover is activated by the naive approach until the network expires.

Since no trap coverage based scheduling algorithm has been proposed before, *Naive-Trap* which guarantees D-trap coverage is considered to be the state-of-the-art solution.

Simulations are performed for these four algorithms under the same setting. We assign weights to sensors at the beginning of each time slot and treat each intersection point as a target in *Greedy-MSC Heuristic*. *DSC* is an area coverage protocol to ensure full coverage. We compare trap coverage with full coverage in the simulations because there is no existing work of conducting experiments to verify the significant improvement of energy consumption and lifetime under trap coverage in WSNs of large scale.

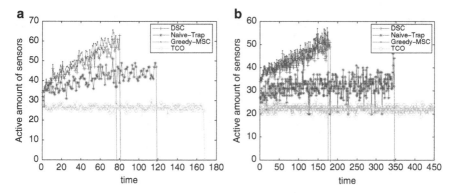

Fig. 3.12 Active amount of sensors vs. time slot. (**a**) $E = 20$, $D = 15$, $N = 300$. (**b**) $E = 30$, $D = 25$, $N = 400$

Since we assume that inactive sensors do not consume any energy, the number of active sensors per time slot denotes the energy consumption. We conduct simulations to compare the active number of sensors during the lifetime running by these algorithms. The results in Fig. 3.12a, b suggest that the energy consumption of our algorithm is the lowest, which may lead to a longer lifetime. In order to balance the energy consumption of sensors, sensors with more residual energy are activated with higher priority. Results in Fig. 3.13a, b show the average residual energy of activated sensors by these algorithms, which demonstrates that TCO always activates sensors with higher residual energy. We also illustrate the coverage of these algorithms during a time slot in Fig. 3.14, where blue circles denotes the sensing region of activated sensors. Our algorithm apparently activates less sensors to provide required quality of *trap coverage* compared with *Naive-Trap* and the full coverage approach (here we employ DSC) according to Fig. 3.14. We can learn that trap coverage model is an energy-efficient model since its energy consumption per slot is only half of that of full coverage model while the diameter of coverage hole is constrained to be below the diameter of sensor's sensing region, which might be acceptable in many cases. We have two observations about the results. Firstly, Naive-Trap always picks up active sensors without backtracking, while TCO finds

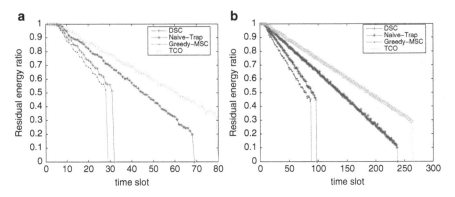

Fig. 3.13 Average residual energy of activated sensors vs. time slot. (a) $E = 20$, $D = 15$, $N = 300$. (b) $E = 30$, $D = 25$, $N = 400$

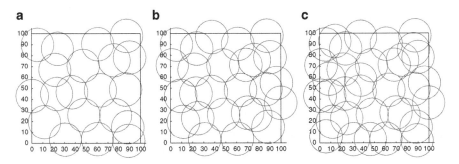

Fig. 3.14 Coverage during a time slot of Naive-Trap, TCO and full coverage approach in RoI. $N = 300$, $E = 20$, $D = 25$. *Circles* denote the sensing region of activated sensors. (a) TCO (20 active sensors). (b) Naive-Trap (26 active sensors). (c) Full coverage (38 active sensors)

a minimum weight sensor cover at the beginning and then removes the redundant sensors, which means TCO determines active sensor set globally and thus more efficiently. Secondly, TCO considers the effect of each sensor on the diameter of coverage hole directly. $D_\psi(i)$ of sensor i as the upper bound of increment of coverage hole's diameter is taken into consideration, which can significantly reduce the amount of activated sensors. The importance of $D_\psi(i)$ is validated in Fig. 3.4. We define $G(i)$ as the normalized weight in TCO to trade off between upper bounds of increment and weights of sensors, and remove sensors from initial sensor set based on $G(i)$. Therefore, our algorithm outperforms *Naive-Trap*.

Fig. 3.15 Lifetimes of
Naive-Trap, TCO, DSC and
Greedy-MSC Heuristic. (**a**)
Lifetimes vs. N, $E = 10$,
$D = 15$. (**b**) Lifetimes vs. E,
$N = 300$, $D = 15$. (**c**)
Lifetimes vs. D, $N = 300$,
$E = 20$

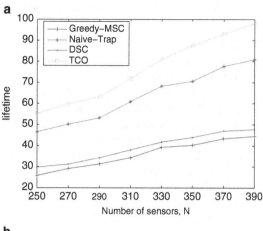

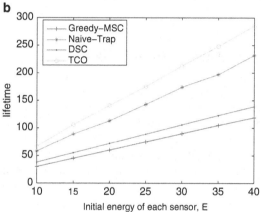

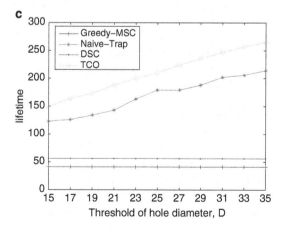

3.6.3 Lifetime Performance Evaluation

The lifetimes achieved by *Naive-Trap*, *TCO*, *DSC* and *Greedy-MSC Heuristic* versus different scenarios are plotted in Fig. 3.15a, b. The lifetime is lengthened if the number of deployed sensors N or the initial energy of each sensor E increases. The plots suggest that *TCO* always has a better performance of longevity compared with *Naive-Trap* in different scenarios. That is because TCO always has lower energy consumption and activates sensors with higher residual energy, which is shown in aforementioned plots. The simulation results also prove that trap coverage can extend the network lifetime significantly. We vary the diameter threshold D to compare the lifetimes achieved by *Naive-Trap*, *TCO*, *DSC* and *Greedy-MSC Heuristic* in Fig. 3.15c. The network lifetime increases if we allow a larger D. Since *Greedy-MSC Heuristic* and *DSC* has nothing to do with the diameter D, the lifetime remains unchanged when D varies.

3.6.4 Communication Cost

We evaluate the communication cost of localized trap cover protocol even though we do not consider communication cost in the simulation of lifetime. We denote the average communication cost by the average amount of messages that a sensor needs to send in the protocol. Communication collision and message dropping are not considered in this simulation. Since LTCP adopts DSC as its first step, the average communication costs of DSC are also included in LTCP. We also introduce an estimate amount of messages $(2 + 4\pi r^2 \rho')\pi(r + D)^2 \rho'$ which is calculated in Sect. 3.5.2 as a reference value for the average messages of LTCP except DSC to evaluate the formula. r is set to be 15. We conduct simulations on the average messages per sensor sent in DSC for comparison. The amount of messages of each sensor in localized trap cover protocol are plotted in Fig. 3.16a, b. The amount of messages increases when D increases in Fig. 3.16a as we analyzed. The average communication costs of DSC depend on the density of sensors in the field. We can see that the communication overhead of LTCP except DSC is low. That is because the communication overhead of LTCP except DSC are only cost by the sensors in the minimum weight sensor cover whose amount is limited. It is trivial when the overhead is shared by all sensors.

As shown in Fig. 3.16b, the average communication costs of LTCP do not vary when the amount of sensors increases. The communication costs of LTCP except DSC will not vary when the amount of sensors varies since ρ', the density of sensors which are activated to provide initial full coverage, will not be influenced by the total amount of sensors N. Besides this, the communication costs of DSC increase when N increases, while the amount of all sensors also increases to share the communication costs of LTCP. These two factors neutralize, so the communication costs of LTCP will not vary significantly when the amount of all sensors varies.

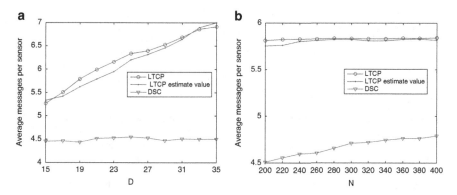

Fig. 3.16 Average amount of messages per sensor. (**a**) Average amount of messages vs. D, $N = 200$. (**b**) Average amount of messages vs. N, $D = 15$

Also, the estimate value is close to the actual amount of messages, which validates our analysis in Sect. 3.5.2.

3.7 Conclusions

In this chapter, we have investigated the problem of trap coverage in WSNs. *Minimum Weight Trap Cover Problem* is formulated to schedule the activation of sensors in WSNs under the model of trap coverage. We always activate the minimum weight trap cover successively at each time slot to balance the energy consumption of each sensor so that the longevity of networks is ensured. A novel algorithm is proposed to tackle with the problem based on trap coverage which is shown in simulation results to have better performance than the state-of-the-art approach. We also design a localized protocol to implement our algorithm in WSN which is proved to have the same performance as the global solution. The performance of Minimum Weight Trap Coverage we find is proved to be at most $O(\rho)$ times of the optimal solution, where ρ is the density of sensor nodes in the region.

References

1. J. Ko, C. Lu, M. Srivastava, J. Stankovic, A. Terzis, and M. Welsh. Wireless sensor networks for healthcare. *Proceedings of the IEEE*, 98(11):1947–1960, 2010.
2. X. Cao, J. Chen, Y. Zhang, and Y. Sun. Development of an integrated wireless sensor network micro-environment monitoring system. *ISA Transactions (Elsevier)*, 47(3):247–255, 2008.
3. S. Zahedi, M. Srivastava, C. Bisdikian, and L. Kaplan. Quality tradeoffs in object tracking with duty-cycled sensor networks. In *Proceedings of IEEE Real-Time Systems Symposium (RTSS)*, 2010.

4. Y. Liu and W. Liang. Approximate coverage in wireless sensor networks. In *IEEE International Conference on Local Computer Networks (LCN)*, 2005.

5. M. Cardei, T. Thai, Y. Li, and W. Wu. Energy-efficient target coverage in wireless sensor networks. In *Proceedings of IEEE International Conference on Computer Communications (INFOCOM)*, 2005.

6. S. He, J. Chen, and Y. Sun. Coverage and connectivity in duty-cycled wireless sensor network for event monitoring. *IEEE Transactions on Parallel and Distributed Systems*, 23(3):475–482, 2012.

7. S. Slijepcevic and M. Potkonjak. Power efficient organization of wireless sensor networks. In *IEEE International Conference on Communications (ICC)*, pages 472–476, 2001.

8. J. Chen, J. Li, S. He, Y. Sun, and H. Chen. Energy-efficient coverage based on probabilistic sensing model in wireless sensor networks. *IEEE Communication Letters*, 14(9):833–835, 2010.

9. J. Jeong, Y. Gu, T. He, and D. Du. Visa: Virtual scanning algorithm for dynamic protection of road networks. In *Proceedings of IEEE International Conference on Computer Communications (INFOCOM)*, 2009.

10. G. Wang, G. Cao, and T. Porta. Movement-assisted sensor deployment. In *Proceedings of IEEE International Conference on Computer Communications (INFOCOM)*, 2004.

11. B. Liu and D. Towsley. A study of the coverage of large-scale sensor networks. In *IEEE International Conference on Mobile Ad-hoc and Sensor Systems (MASS)*, 2004.

12. P. Balister, Z. Zheng, S. Kumar, and P. Sinha. Trap coverage: Allowing coverage holes of bounded diameter in wireless sensor networks. In *Proceedings of IEEE International Conference on Computer Communications (INFOCOM)*, 2009.

13. J. Li, J. Chen, S. He, T. He, Y. Gu, and Y. Sun. On energy-efficient trap coverage in wireless sensor networks. In *Proceedings of IEEE Real-Time Systems Symposium (RTSS)*, 2011.

14. G. Mao, B. Fidan, and B. Anderson. Wireless sensor network localization techniques. *Computer Networks*, 51(10):2529–2553, 2007.

15. N. Patwari, J. Ash, S. Kyperountas, A. Hero, R. Moses, and N. Correal. Locating the nodes: cooperative localization in wireless sensor networks. *IEEE Signal Processing Magazine*, 22(4):54–69, 2005.

16. J. Hwang, T. He, and Y. Kim. Exploring in-situ sensing irregularity in wireless sensor networks. In *Proceedings of the International Conference on Embedded Networked Sensor Systems (SenSys)*, 2007.

17. T. Yan, Y. Gu, T. He, and J. Stankovic. Design and optimization of distributed sensing coverage in wireless sensor networks. *ACM Transactions on Embedded Computing Systems*, 7(3):1–40, 2008.

18. X. Wang, G. Xing, Y. Zhang, C. Lu, R. Pless, and C. Gill. Integrated coverage and connectivity configuration in wireless sensor networks. In *Proceedings of the International Conference on Embedded Networked Sensor Systems (SenSys)*, 2003.

19. G. Kasbekar, Y. Bejerano, and S. Sarkar. Lifetime and coverage guarantees through distributed coordinate-free sensor activation. In *Proceedings of the Annual International Conference on Mobile Computing and Networking (MobiCom)*, 2009.

20. S. Yang, F. Dai, M. Cardei, and J. Wu. On multiple point coverage in wireless sensor networks. In *Proceedings of IEEE Conference on Mobile Adhoc and Sensor Systems Conference (MASS)*, 2005.

21. R. M. Karp. Reducibility among combinatorial problems. In *50 Years of Integer Programming 1958-2008*, pages 219–241. Springer Berlin Heidelberg, 2010.

22. P. Berman, G. Calinescu, C. Shah, and A. Zelikovsky. Efficient energy management in sensor networks. In *Ad Hoc and Sensor Network*. Nova Science Publishers, 2005.

23. M. Cardei and D. Du. Improving wireless sensor network lifetime through power aware organization. *Wireless Networks*, 11(3):333–340, 2005.

24. X. Bai, S. Kumar, D. Xuan, Z. Yun, and T. Lai. Deploying wireless sensors to achieve both coverage and connectivity. In *Proceedings of the ACM International Symposium on Mobile Ad Hoc Networking and Computing (MobiHoc)*, 2006.
25. V. Chvatal. A greedy heuristic for the set-covering problem. *Mathematics of Operations Research*, 4(3):233–235, 1979.
26. T. Cormen, C. Leiserson, R. Rivest, and C. Stein. Data structures for disjoint sets. In *Introduction to Algorithms*, pages 498–524. MIT Press and McGraw-Hill, 2001.

Chapter 4
Trapping Mobile Intruders in Sensor Networks

4.1 Introduction

As discussed in Chap. 3, trap coverage has been proposed as a promising alternative to full coverage for applications such as intrusion detection, target location and environment monitoring, especially in the scenario of mobile targets detection [1]. In full coverage, sensors are deployed over a region of interest (RoI) such that every point/location in the region is within the sensing range of at least one sensor (or k sensors if k-full coverage is desired). If the RoI is a wide area, as is in many applications, full coverage may be too conservative and expensive to achieve. Hence arises the concept of trap coverage [1], which allows "uncovered holes" in the RoI as long as they are no larger than a pre-specified size. Mobile intruders (targets) in uncovered holes are presumably undetected, but once they move out of a hole, they will be detected. That is sufficient for many applications, and different applications allow different hole sizes. Despite the loss in sensing performance, trap coverage has been shown to be much more energy-efficient than full coverage [2]. It provides sufficient sensing performance with much fewer sensors than needed for full coverage. To explore the fundamental metric of trap coverage straightly, existing literatures [1, 2] have made assumption of disc sensing model with circle sensing region to simplify the problem. A sensor with a disc sensing model can detect any target within a certain range (called its sensing range) with probability 1, but cannot detect anything beyond the range.

Recently, it has been discovered that sensors in practical scenarios can be modeled more accurately with the probabilistic model [3–7]. In this model, a sensor is able to detect a target at a distance d away from the sensor with probability $\lambda(d)$, where

$$\lambda(d) = \begin{cases} 1, & \text{if } d \leq d_1 \\ g(d), & \text{if } d_1 < d \leq d_2 \\ 0, & \text{if } d_2 < d \end{cases} \qquad (4.1)$$

S. He et al., *Energy-Efficient Area Coverage for Intruder Detection in Sensor Networks*, 69
SpringerBriefs in Computer Science, DOI 10.1007/978-3-319-04648-8_4,
© The Author(s) 2014

for some decreasing function $g(d)$ with $0 \leq g(d) \leq 1$. Various functions have been proposed for $g(d)$. Note that when $d_1 = d_2$, the probabilistic model degenerates to the disc sensing model. Actually, the concept of uncovered holes is no longer applicable in probabilistic sensing model and we can not guarantee to trap a mobile target by restricting the size of uncovered holes any more. Yet no existing literatures are concerned with trapping mobile targets based on probabilistic sensing model.

The speeds of mobile targets should be quite different in a variety of scenarios. Sensors always perform sensing task discretely, thus targets moving faster should have a greater displacement before being detected. It is not considered in the disc sensing model. However, the maximum speed of mobile target is necessary to be considered when trapping the target. For example, sensor networks designed for monitoring humans should be different from that for monitoring fighters in a battle field since human and fighter have different maximum moving speeds.

To extend the concept of trap coverage into a real large-scale WSN, we investigate some fundamental problems of trapping mobile targets with probabilistic sensors. We define probabilistic trap coverage in this chapter. The quality of probabilistic trap coverage can be measured by two parameters: (D, ϵ), where D is a distance and ϵ is a probability. A sensor network provides (D, ϵ)-trap coverage (see an illustration in Fig. 4.1) to a RoI if any mobile target in the region is guaranteed to be detected with probability at least ϵ should it move for a displacement of D from its original position. (The notion of (D, ϵ)-trap coverage will be more precisely defined in Sect. 4.2.) By guaranteeing (D, ϵ)-trap coverage, each potential undetected mobile target in the RoI is trapped in a circle of diameter less than D.

As mentioned above, in the scenario of trapping mobile targets, the moving speed is an important factor. Intuitively, it is harder to detect a target if it moves faster. Probabilistic trap coverage is much more effective than traditional trap coverage to describe the detection of mobile target with various speeds. In this chapter, the detection probability of mobile targets with various moving speeds is analyzed theoretically to establish the foundation of probabilistic trap coverage. Analyzing the detection probability of a mobile target is nontrivial because of its arbitrary moving path. It is rather difficult to determine the probability for infinite possible paths. We successfully find a lower bound of detection probabilities among all possible paths by developing a theory of **circular graph**, with the help of which we propose a *Circular Graph Algorithm* to check whether a given sensor network can provide (D, ϵ)-trap coverage or not. The theory of circular graph has potential to be applied to other coverage problems as well. Indeed, we have used it to solve a closed-belt barrier coverage problem as described in [8]; we will also report this result in this chapter.

Assume that sensors have been deployed over a RoI for trap coverage. We are also interested in practical energy-efficient issues such as scheduling sensors to provide (D, ϵ)-trap coverage to the RoI, while attempting to maximize the sensor network's lifetime. An efficient localized sleep-wakeup protocol, *Probabilistic Trap Coverage Protocol* is designed to provide desired probabilistic trap coverage while maximizing the network lifetime. The lower bound of lifetime acquired by the protocol is nearly 1/2 of the optimum lifetime.

Our main contributions can be summarized as below [9]:

1. We extend the concept of trap coverage into a realistic model and analyze the detection probability of mobile targets with various moving speeds travelling along arbitrary path in RoI theoretically, based on which probabilistic trap coverage is defined.
2. We develop a theory of circular graph for the problem of probabilistic trap coverage. The theory is interesting in its own right, and with its help, we develop an algorithm that determines whether a sensor network can provide (D, ϵ)-trap coverage.
3. Even though our theory of circular graph was developed to solve the problem of probabilistic trap coverage, it can be used to solve a difficult problem discussed in [8]: Does a sensor network over a closed/circular belt provide k-barrier coverage? We provide a simple solution to that problem.
4. We formulate and study the problem of scheduling the activation of sensors energy-efficiently while providing desired probabilistic trap coverage in a large scale WSN. We design an efficient localized protocol to solve the problem. The lower bound of lifetime acquired by the protocol is proven to be nearly 1/2 of the optimum lifetime. Extensive simulations are conducted to validate the efficiency of the protocol.

The rest of the chapter is organized as follows. We present the formulation of *Maximum Trap Network Lifetime Problem* in Sect. 4.2. The theoretical analysis is conducted in Sect. 4.3. Section 4.4 presents the details of protocol design and extensive simulations are performed to verify the effectiveness of protocol in Sect. 4.5. The chapter is concluded in Sect. 4.6.

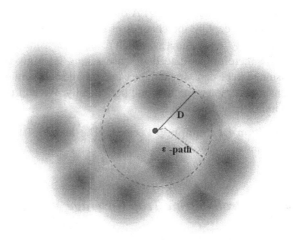

Fig. 4.1 An example of probabilistic trap coverage

4.2 Preliminary and Problem Statement

4.2.1 Network Model

We consider a large scale WSN consisting of a set of sensors deployed in a vast two-dimension RoI A. We assume that sensors are randomly deployed in the region A and there are no two sensors deployed at the same location.

The location of each sensor is assumed to be known, either by GPS or other localization techniques [10, 11]. Each sensor can only communicate with other sensors within its *transmission range R*. We adopt the probabilistic sensing model. The detection probability is assumed to be a continuously decreasing function of distance. Several empirical detection probability formulas have been proposed, e.g., the exponential attenuation probabilistic model [12] and the cubic attenuation model [4]. Most existing probabilistic sensing models (including exponential attenuation and cubic attenuation models) use a function which is concave with a positive second-order derivative. Since these empirical detection probability formulas are estimated from realistic data, the following assumption, which we will need in this chapter, is not unrealistic or uncommon. In this chapter, we assume such properties, as formally stated in the following.

Assumption 4.1. *The detection probability function is concave in the interval where the probability attenuates as the sensing distance increases, whose second order derivative is positive.*

For example, the sensing function can be given by Eq. (4.1).

The majority of sensors perform sensing task discretely based on the following facts that (i) Sensors with on board micro processor units such as Mica2 and Imote2 sample in a certain rate; (ii) A directional sensor, maybe a camera or a radar, has to rotate to capture the changes of surroundings; (iii) Sensors may be set to sleep periodically to save energy, which results in sensing intermittently.

We therefore assume that sensors *sense* at a given sampling rate. The rate is assumed to be constant and the same for all sensors in this chapter. Sensor performs sensing task intermittently and periodically. We will use the term *sampling period* to indicate the following.

Definition 4.1 (Sampling Period). The length of time between two consecutive samplings.

4.2.2 Probabilistic Trap Coverage Model

To trap mobile targets within a limited area in the RoI, we define probabilistic trap coverage. Firstly, we will introduce the definition of ϵ-path.

Definition 4.2 (ϵ-path and its displacement). An ϵ-**path** is a path along which a mobile target can travel with a detection probability lower than ϵ. Its **displacement** is the Euclidean distance between its start point and end point.

If p^l is the detection probability of ϵ-path l, we have

$$p^l < \epsilon \tag{4.2}$$

The notion of ϵ-path describes a set of paths with detection probability lower than the threshold ϵ. We will discuss how to figure out the detection probability of a path in Sect. 4.3. Based on the definition of ϵ-path, we give a strict definition of *probabilistic trap coverage*. Probabilistic trap coverage restricts the maximum displacement among all ϵ-paths in RoI.

Definition 4.3 ((D, ϵ)-trap coverage). A set of sensors C provides (D, ϵ)-trap coverage to RoI A if the displacement of every ϵ-path in A is not greater than D. A point X is said to be (D, ϵ)-trap covered if the displacement of every ϵ-path in A started from X is not greater than D.

Apparently if all points in RoI A are (D, ϵ)-trap covered, A is (D, ϵ)-trap covered. If RoI is (D, ϵ)-trap covered, a mobile target will be detected with a probability not less than ϵ if it moves with a displacement of D at most. In fact, if we set $\epsilon = 1$ and assume the detection probability can only be 0 and 1, probabilistic trap coverage turns into traditional trap coverage. In this way, traditional trap coverage is only a special case of probabilistic trap coverage.

4.2.3 Problem Statement

The operation time of each individual sensor is divided into (time) slots. We assume that each sensor has an initial energy of E units and consumes 1 unit per slot if it is active. The inactive sensor consumes negligible energy. Only sensor with residual energy no less than 1 unit can be activated.

Suppose that a WSN of N sensors has been deployed in RoI A. We are interested in energy-efficiently scheduling the sensors so as to provide (D, ϵ)-trap coverage to the region while maximizing the lifetime of the WSN.

Random deployment may lead to much redundancy of sensors in the monitored region [13]. To improve the energy-efficiency and maximize the lifetime of WSN, the sensors switch between sleep mode and active mode according to the scheduling strategy. It is a challenging task to guarantee (D, ϵ)-trap coverage energy efficiently. We formulate the problem below. It is desired that the scheduling algorithm is distributed and localized.

Maximum Trap Network Lifetime Problem: Given a region A, a set of homogeneous sensors $\{1, 2, \cdots, N\}$ with initial energy supply E, sampling period T, transmission radius R and detection probability function. The location of each sensor i is known. We wish to find a schedule that maximizes the lifetime of network while always providing (D, ϵ)-trap coverage to region A, where D and ϵ are parameters set by applications.

Given the parameters, there exists a family of sensor sets $\aleph$ which are $C_k, k = 1, 2, \cdots$, and each set C_k in $\aleph$ can provide probabilistic trap coverage. We assume t_k is the activated time slots of set C_k and t_k can only be non-negative integers. If sensor i belongs to set C_k, then $x_{ik} = 1$; otherwise, $x_{ik} = 0$. The problem can be formally formulated as follows.

$$
\begin{aligned}
\max \quad & \sum_{C_k \in \aleph} t_k \\
s.t. \quad & \sum_{C_k \in \aleph} t_k x_{ik} \leq E, i = 1, 2, \cdots, N
\end{aligned}
\tag{4.3}
$$

Note that if we set $\epsilon = 1$ and assume the detection probability can only be 0 and 1, the problem degenerates into the traditional trap coverage problem with the assumption of disc sensing model. It indicates that to maximize the lifetime of sensor network while guaranteeing traditional trap coverage is a special case of our problem.

4.3 Probabilistic Trap Coverage

4.3.1 Detection Gain

In this section, we establish some fundamental results for probabilistic trap coverage. Let C be a set of active sensors over a region, and l denote a path in the region. If a mobile target moves along l from one end to the other, what is the probability p of the target being detected by at least one sensor? Since sensors detect (sample) at discrete times, the probability p depends on how the mobile target moves—its maximum speed and how the speed varies. In this section, given sensors sampling period T and the target's maximum speed V_{max}, we compute a lower bound on the above mentioned probability p. We start with one sensor, then two sensors, and then any set of sensors. Actually, We have analyzed the scenario of one sensor and two sensors in [14]. The results are summarized below in Sects. 4.3.1 and 4.3.2.

Let i denote a sensor. Let $p_i(j)$ be the detection probability of point j by sensor i. As the target moves along the path, let Ω_i^l denote the set of points/locations at which sensor i performs the detection function. Then, p_i^l, the probability of the target detected by i, satisfies

$$
1 - p_i^l = \prod_{j \in \Omega_i^l} [1 - p_i(j)]
\tag{4.4}
$$

Now consider a set of sensors C. The probability p_C^l (or simply p^l) that at least one sensor in C detects the mobile target, satisfies

$$1 - p^l = \prod_{i \in C} \prod_{j \in \Omega_i^l} [1 - p_i(j)] \tag{4.5}$$

where $1 - p^l$ is the probability that the mobile target is *not* detected by any sensor in C. Linearize Eq. (4.5) by taking the natural logarithm on both sides, we have

$$\ln(1 - p^l) = \sum_{i \in C} \sum_{j \in \Omega_i^l} \ln[1 - p_i(j)] \tag{4.6}$$

Note that $-\ln(1 - p^l)$ is a strictly increasing function of p^l. In subsequent sections we will need to compare different probabilities, say p^l and p^t. Instead of comparing p^l and p^t directly, we will compare $-\ln(1 - p^l)$ and $-\ln(1 - p^t)$, because the latter are easier to compute due to Eq. (4.6). In view of its importance, we make the following definition.

Definition 4.4 (Detection Gain). If p is a detection probability, the quantity $\phi = -\ln(1 - p) = \ln \frac{1}{1-p}$ is the detection gain corresponding to p.

If a path l has a larger detection probability than another path t, then l has a larger detection gain than t, and vice versa. We will use detection gain as a convenient substitute of detection probability.

Now, if ϵ is a threshold (any threshold) of detection probability, then in the domain of detection gain, we have a corresponding threshold $-\ln(1 - \epsilon)$, as defined below.

Definition 4.5 (Aggregate Detection Threshold). If ϵ is a detection probability threshold, then the corresponding aggregate detection threshold is $\varphi(\epsilon) = \ln \frac{1}{1-\epsilon}$.

If l is an ϵ-path, then $p^l < \epsilon$, or equivalently, l's detection gain is less than the aggregate detection threshold $\varphi(\epsilon)$:

$$\begin{aligned}
&\ln(1 - p^l) > \ln(1 - \epsilon) \\
&\Rightarrow \sum_{i \in C} \sum_{j \in \Omega_i^l} \ln \frac{1}{1 - p_i(j)} < \ln \frac{1}{1-\epsilon}
\end{aligned} \tag{4.7}$$

The detection gain of path l is denoted by ϕ^l and the detection gain of point j by sensor i is denoted by $\phi_i(j)$. $\phi^l = \ln \frac{1}{1-p^l}$ and $\phi_i(j) = \ln \frac{1}{1-p_i(j)}$.

$$\phi^l = \sum_{i \in C} \sum_{j \in \Omega_i^l} \phi_i(j) < \varphi(\epsilon) \tag{4.8}$$

Note that if $p_i(j) = 1$, $\phi_i(j) = \infty$, which indicates that the detection probability is always sufficient for the required threshold no matter how large ϵ is.

4.3.2 *Impact of Maximum Speed*

Continuing our discussion in the preceding section, we wish to find a lower bound
on all paths' detection gains. Consider first the case of two sensors located at points
a and b. It can be proven under Assumption 4.1 that of all (sufficiently long) paths
passing between a and b the one with minimum detection gain is the line along
the perpendicular bisector of the segment between a and b, with the target moving
constantly at its maximum speed, and with the two sensors happening to sample at
the same time (i.e., $\Omega_a^l = \Omega_b^l$, which we will simply denote as Ω^l). (The proof is
referred in [14].) Figure 4.2 illustrates the situation. This minimum detection gain
can be computed (estimated) as follows.

Fig. 4.2 The path in worst
detection case

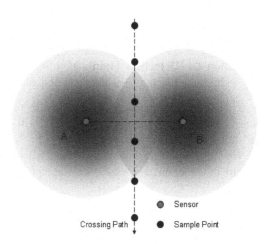

Sensor

Crossing Path   Sample Point

Given the sampling period T and the maximum speed of intruders V_{max}, the
distance between each two detection points along the perpendicular bisector is
$T * V_{max}$. Points which are too far away from these two sensors with negligible
detection gain are not considered in Ω^l. What distance is considered too far away
depends on the value of d_2 in Eq. (4.1). Since we are interested in a lower bound on
detection gain, excluding some points from Ω^l still yields a lower bound. Further,
when we calculate the detection gain—actually a lower bound—across the area
between two sensors in a network, we use the Voronoi Graph and only sum up
the detection gains of sampling points lying on the Voronoi edge between the two
sensors. (Consecutive sampling points are spaced by a distance of $T * V_{max}$.) For
example, the detection gain of sensors A and B is calculated by the sampling
points which lie on their Voronoi edge (the dotted line between A and B) as shown
in Fig. 4.3.

Note that we only sum up the detection gain of the nearest two sensors on the path
along Voronoi path. We here explain why it is the lower bound of detection gain.
Consider an actual intruder moves along the crossing path l as shown in Fig. 4.4,
the detection gain of all sensors should be considered. For example, it has to move
across A and B, thus has a detection gain greater than the detection gain calculated

on Voronoi edge between A and B. However, when it is moving across A and B, it is also on the way across B and C. The detection gain of C therefore should also be added. We divide the detection process among A, B and C into two separated process and each should have a greater detection than that on the path along the perpendicular bisector. In this view, the sum of the detection gain on the path along Voronoi path by the nearest two sensors is the lower bound. Above all, the analysis of Voronoi graph is dividing the detection process at the same time into different parts in the Voronoi graph which seems to be sequential.

Fig. 4.3 The Voronoi graph of multiple sensors

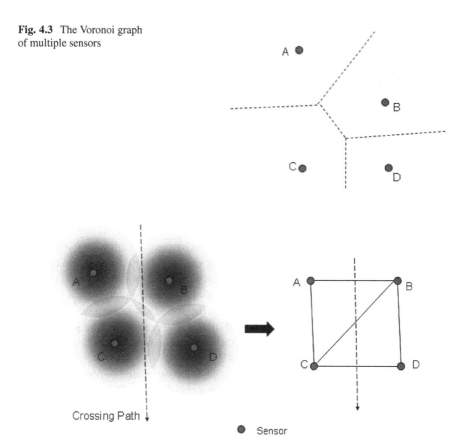

Fig. 4.4 The illustration of multiple sensors scenario

4.3.3 Circular Graph

Suppose there is a set of sensors deployed over a region. We can easily obtain a lower bound on the detection gain of a target moving along any path. (For simplicity, we sometimes call it minimum detection gain.) Given the linearity of detection gain, we can just sum up the detection gain of each pair of sensors. For example, if we want to calculate the minimum detection gain to move in the sensor network through the arrow path in Fig. 4.4, we just sum up the minimum detection gain of each pair of connected sensors whose link intersects with the path l.

Now consider a target located at a point X. If the target moves to a new location, say Y, which is a distance of D away from X, what would be the minimum detection gain? If for every point X in the region that the minimum detection gain is greater than the aggregate detection threshold $\varphi(\epsilon)$, then we know the sensor network provides (D, ϵ)-trap coverage.

To answer the above question, we construct a *circular graph* as illustrated in Fig. 4.5 to characterize the feature of probabilistic trap detection. It is about the relationship of sensors around the point X. In the circular graph, nodes represent sensors and there is an edge between two sensors iff they are neighbors in the sensors Voronoi graph. The capacity of the edge is the minimum detection gain of two corresponding sensors. The circular graph is only composed of sensors within distance D from point X (including the distance equals to D). If a target located at X moves out the circular graph, the detection gain is the aggregate capacities of edges which intersect with the moving path. According to the definition of (D, ϵ)-trap coverage, we have Theorem 4.1.

Theorem 4.1. *Point X is (D, ϵ)-trap covered, if a target at point X can not escape out of its circular graph through an ϵ-path, i.e, a path with detection gain lower than $\varphi(\epsilon)$.*

Thus, we need to determine whether there exists any ϵ-path along which the target can move out the circular graph. To that end, we define a *cut* in the circular graph as a set of edges without which the graph will not be closed (or circular). The edges which intersect with a path escaping out of a circular graph form a *cut*. (For instance, in Fig. 4.5, the two edges intersecting the dotted arrow are a cut.) The total capacity of the cut provides a lower bound on the aggregate detection gain of that path. If there exists no ϵ-path, the capacity of any cut should be greater than the aggregate detection threshold $\varphi(\epsilon)$.

A *circular flow* in a circular graph is defined as a flow around the center point X. The flow has no source and no sink, and forms a circular loop. It is like water flows in a closed pipeline network and the water just flows in a circulating way around the center point X with no water leaks. We have the following "max-flow min-cut" theorem for circular graphs.

Theorem 4.2. *The minimum capacity of a cut equals the maximum circular flow of a circular graph.*

Theorem 4.2 can be been proved in a way similar to that of the well known Max-flow Min-cut Theorem in graph theory. All circular flows must not be greater than any cut due to the limit of capacity. But there must exist a cut K which equals the maximum circular flow in the graph, otherwise we can find a circulating route in the graph around point X with unoccupied capacity to increase the maximum circular flow. Thus, the maximum circular flow must equal the minimum aggregate capacity of edges in cut, i.e., the minimum cut.

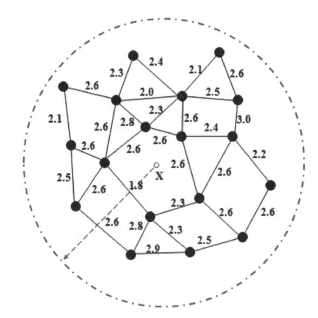

Fig. 4.5 An example of circular graph. The path with minimum detection gain to move as a displacement of D is marked by splash line

Therefore, a point X is (D, ϵ)-trap covered if its circular graph has a maximum circular flow greater than $\varphi(\epsilon)$. We now present our approach to calculating the maximum circular flow of a circular graph.

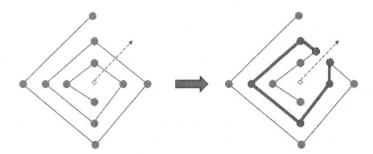

Fig. 4.6 A counterexample to the super-virtual-node approach. The maximum circular flow of the left circular graph is zero, but a flow (in bold edges) exists in the right cut graph

At first glance, it seems we can cut the circular graph as illustrated in Fig. 4.6 (left) and add two super virtual nodes as source and sink, thereby yielding a flow network (Fig. 4.6 right); then calculate the maximum flow of the flow network, which "should" equal the maximum circular flow of the original circular graph. This argument is not valid. Figure 4.6 gives a counterexample. The left circular

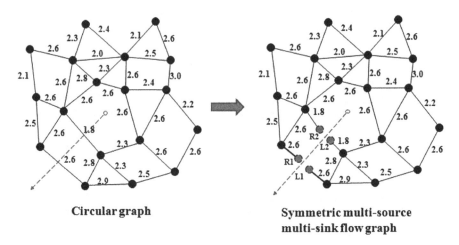

Circular graph   **Symmetric multi-source multi-sink flow graph**

Fig. 4.7 The approach to transforming a circular graph into a symmetric multi-source multi-sink flow graph

graph is not closed and its maximum circular flow is zero, i.e., the center point is not trap-covered, but the right graph has a flow (in bold lines) between the source and the sink. Thus, that is not a correct way to calculate the maximum circular flow.

To solve this problem, we first transform the circular graph (around a point X) into a *symmetric multi-source multi-sink flow graph*, in which sources and sinks are in pairs with the same edge capacity. The circular graph looks like a ring or bracelet. We arbitrarily choose a path emitting from X to cut the ring as illustrated in Fig. 4.7 (left). Then, we add pairs of virtual nodes to take place of the cut edges as shown in Fig. 4.7 (right). That is, each cut edge is transformed into a pair of virtual nodes—a *left* node and a *right* node. Each virtual node connects to its neighboring node with an edge of the same capacity as the original (cut) edge. The resulting graph is a symmetric multi-source multi-sink flow graph, with added left virtual nodes serving as sources and right virtual nodes as sinks.

A *restricted flow* in a symmetric multi-source multi-sink flow graph is an ordinary network flow from left nodes (sources) to right nodes (sinks) with the additional property that the out-flow of each left node equals the in-flow of its corresponding right node. The following theorem is easy to prove.

Theorem 4.3. *The maximum circular flow of a circular graph equals the maximum restricted flow of the transformed symmetric multi-source multi-sink flow graph.*

It is a bit different to calculate a maximum "restricted" flow than a traditional (non-restricted) maximum multi-source multi-sink flow. In the latter case, one simply adds a super virtual source and a super virtual sink connected to all the source nodes and all the sink nodes, respectively, and then computes a maximum flow from the super source node to the super sink node. This approach does not work for restricted maximum flows because of the restriction of restricted flows.

Instead, we use a modified Edmonds-Karp algorithm. The original Edmonds-Karp algorithm computes a maximum flow (in a single-source single-sink flow graph) by repeatedly finding a shortest *augmenting path*[1] from the source node to the sink node. Our modified Edmonds-Karp algorithm (see the next paragraph for more details) computes a maximum "restricted" flow (in a symmetric multi-source multi-sink flow graph) by repeatedly finding a shortest augmenting path from a source node to its corresponding sink node, until there exists no more augmenting path between any pair of source and sink in the *residual graph*. The modified Edmonds-Karp algorithm generates a maximum restricted flow; its proof of correctness is very similar to that of the original Edmonds-Karp algorithm [15] and thus omitted here.

For completeness, we describe here our modified Edmonds-Karp algorithm in a little more detail. Generally, we denote each undirected edge between u and v in the symmetric multi-source multi-sink flow graph by two directed edge from u to v and from v to u with the same capacities to transform the undirected graph into an equivalent directed graph so that we can calculate the maximum restricted flow. Let $G(E_d, V)$ be the equivalent directed graph, where E_d is the set of edges and V is the set of nodes. Also, for each edge from u to v, let $c(u, v)$ be the capacity and $f(u, v)$ be the flow. Mark the set of left virtual nodes as the source set and the set of right virtual nodes as the sink set. Apparently the source set and the sink set have no common element. Each node s in the source set has a corresponding node s' in the sink set. We want to find the maximum aggregate flow from each node in the source set to its corresponding node in the sink set while $f(u, v)$ should not be greater than $c(u, v)$. The residual graph $G_f(V, E_f)$ is defined to be the network with capacity $c_f(u, v) = c(u, v) - f(u, v)$ and no flow.

Modified Edmonds-Karp Algorithm:

   Input: Symmetric multi-source multi-sink flow graph G
   Output: Maximum restricted flow F

1. Denote source set by S and sink set by S'. Set $f(u, v)$ to be zero for all edges.
2. While there is a shortest path p from a node $s \in S$ to its $s' \in S'$ in the residual graph G_f, such that $c_f(u, v) > 0$ for all edges in p: Find $c_f(p) = \min\{c_f(u, v)|(u, v) \in p\}$ and for each edge $(u, v) \in p$, set $f(u, v) = f(u, v) + c_f(p)$ and $f(v, u) = f(v, u) - c_f(p)$ and update the residual graph G_f accordingly.
3. Let F be $\sum_{s \in S}[\sum_{(s,u) \in E_d} f(s, u) - \sum_{(u,s) \in E_d} f(u, s)]$ and return F.

Assume there are k pairs of virtual nodes. The time complexity of modified Edmonds-Karp algorithm is $O(k|V||E_d|^2)$. $|V|$ and $|E_d|$ are the amounts of nodes and edges respectively. We can show that each augmenting path can be found in

---

[1]The reader is referred to [15] or [16] for *augmenting path* and *residual graph*.

$O(k|E_d|)$ time, and every time at least one of the $|E_d|$ edges becomes saturated, that the distance along the augmenting path must be longer than last time and that the length is at most $|V|$.

4.3.4 (D, ε)-Trap Coverage

We have discussed how to determine whether a sensor network provides (D, ϵ)-trap coverage to a point. Now we extend the result to the whole region of interest (RoI).

The RoI A consists of infinite points. It is impossible to check them one by one. We also do not intend to grid the area and approximately guarantee (D, ϵ)-trap coverage for A. We will introduce the concept of *faces* to solve the problem.

A node representing a sensor is included in a point X's circular graph only if X is within distance D of that sensor. Define the *valid border* of sensor i to be the circle of radius D centered around the location of i. Sensor i participates in X's circular graph iff X is within i's valid border. Thus, if we divide the RoI into *subregions* by all sensors' valid borders, then all points in a subregion share the same circular graph. Now, we further divide subregions into *faces* by edges of circular graphs. Then, points in a same face not only share the same circular graph, but are also inside the same cell of the circular graph; their shared circular graph can be cut in the same way and transformed into the same symmetric multiple-source multiple-sink flow graph. (If two points share a same circular graph but are in *different* cells, then their transformed symmetric multiple-source multiple-sink flow graphs will be *different*.) As an example, in Fig. 4.8, the region is divided into thirteen faces, if not considering the area out of the valid borders of all sensors.

Fig. 4.8 An example of face division

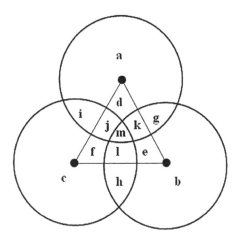

Thus, if any point in the face is (D, ϵ)-trap covered, all points in that face are (D, ϵ)-trap covered. Any point on a face border is covered if the latter's neighbor

---

Algorithm 3 Circular graph algorithm

1: Divide the sensing region into faces according to the location of other sensors in the region.
2: **for** each face **do**
3:     Construct circular graph.
4:     Transform circular graph into symmetric multi-source multi-sink flow graph.
5:     Calculate the maximum restricted flow of symmetric multi-source multi-sink flow graph.
6:     **if** maximum restricted flow is lower than $\varphi(\epsilon)$ **then**
7:         return that the region is not (D, ϵ)-trap covered.
8:     **end if**
9: **end for**
10: return that the region is (D, ϵ)-trap covered.

---

faces are covered, since the point contains no fewer sensors in its circular graph than that of its neighbor faces. Thus, we can guarantee (D, ϵ)-trap coverage to the whole region if all faces are (D, ϵ)-trap covered.

There exists a special case that a point can be detected by surrounding sensors initially without moving. Assume the detection probability is ϵ at the sensing distance d_ϵ (related with sensing function). A face which is totally within the range d_ϵ of a sensor should have already been covered. We will not consider these faces when checking whether the whole region is (D, ϵ)-trap covered. They should have a diameter not greater than $2d_\epsilon$.

Assume there exist $|V|$ sensors and $|E_d|$ edges in a certain circular graph. The time complexity of modified Edmonds-Karp algorithm is $O(k * |V||E_d|^2)$, where k is the pairs of virtual nodes in the symmetric multi-source multi-sink flow graph and we have $k \leq |E_d|$. Thus, if there exist F faces in a circle with a radius of $2D$, the time complexity of Algorithm 3 should be $O(F|V||E_d|^3)$.

We summarize the details in Algorithm 3.

4.3.5 Solving an Open Problem in Barrier Coverage

The above result can be used to solve an open problem in barrier coverage [8]. Consider a sensor network deployed over a belt region. A path crossing the belt region is said to be *k-covered* if it intersects the sensing regions of at least k distinct sensors. A sensor network C provides *k-barrier coverage* over a deployment belt region if all crossing paths through the region are k-covered by C. A fundamental problem in barrier coverage is that, given a sensor deployment over a belt region, how does one determine if it provides k-barrier coverage? In the case of open belts, there is a simple theorem in [8]:

Theorem 4.4. *A sensor network C provides k-barrier coverage [to an open belt] if and only if there exist k node-disjoint paths between the two virtual nodes s and t in the coverage graph of C. (The reader is referred to [8] for the construction of coverage graphs.)*

This theorem has proved to be very useful, having been applied in many subsequent papers on open-belt barrier coverage (e.g., [13, 17, 18]). In the case of closed (donut-shaped) belt regions, unfortunately, there is no such a *simple* theorem. Even though an involved theorem (Theorem 4.2) exists in [8], it "works only for those sensor networks whose coverage graphs can be embedded on a compact surface, but not for arbitrary sensor networks deployed over closed regions" [8]. That theorem, wherever it works, involves deep concepts such as *essential cycles, compact surfaces,* and *embeddable on compact surfaces,* which are not typically familiar to computer scientists. This seems to have dimmed the theorem's usefulness, as evidenced by the fact that it has not been applied by anyone—in contrast to the fact that the corresponding theorem for open-belts has been employed in many papers. Since the publication of [8], researchers have been looking for a *simple* theorem for closed belts such as Theorem 4.4 is for open belts. We provide such a theorem here, applying the results from preceding subsections.

Given a sensor deployment C over a closed belt region. We first construct a coverage graph, in which every node represents a sensor and there is an edge between two nodes if and only if their sensing regions have intersection. All edges have capacity 1. Such a coverage graph is a special *circular graph,* and a corresponding symmetric multi-source multi-sink flow graph can be readily constructed as in Sect. 4.3.3. From the results of Sect. 4.3.3, we immediately have the following simple result:

Theorem 4.5. *A sensor network C provides k-barrier coverage in a closed belt region if and only if the maximum circular flow in the coverage graph of C is greater than or equal to k (or equivalently, iff the maximum restricted flow of the transformed symmetric multi-source multi-sink flow graph is at least k).*

Remark. There is an assumption in [8] (Assumption 4.1) for Theorem 4.4: If the sensing regions of two sensors, D_1 and D_2 have overlap, then $(D_1 \cup D_2) \cap B$ is a connected sub-region in B, where B is the belt region to be considered. Theorem 4.5 needs this assumption, too.

4.4 Localized Protocol

4.4.1 Probabilistic Trap Coverage Protocol

In this section, we propose a localized algorithm called *Probabilistic Trap Coverage Protocol* (PTCP) to guarantee (D, ϵ)-trap coverage and maintain energy efficiency in RoI. The operation time is divided into time slots. Each time slot is divided into two parts: an *initial phase* followed by an *action phase*. Every sensor wakes up at the beginning of each time slot. In the initial phase, they communicate with neighboring sensors and decide whether to stay in active mode or switch to sleep mode. Sensors decide on its mode locally and asynchronously. After the initial phase, during the action phase, if sensors choose to be active, they perform sensing, communication and other tasks; otherwise, they switch to sleep mode to save energy. PTCP runs during the initial phase of each slot.

Sensors contend to sleep to save energy. If too many sensors choose to sleep, the requirement of probabilistic trap coverage will not be met. On the other hand, if too many sensors are active, it will be a waste of resources. Thus, desired is a mechanism to coordinate sensors' decisions. We therefore introduce *priority*. Every sensor has a unique ID. At the beginning of each time slot, each sensor is assigned a priority based on its residual energy and ID. Let pr_i denote the priority of sensor i. We define $pr_i = \{E_i, ID_i\}$ where E_i is residual energy of sensor i and ID_i is its ID. $pr_i > pr_j$ if, (i) $E_i < E_j$ or (ii) $E_i = E_j$ and $ID_i < ID_j$. If a sensor has a lower priority than its neighbors, it has to make decision after the sensors with higher priority. In this way, sensors with less residual energy have the higher priority to switch to sleep mode and leave sensors with more residual energy to perform sensing tasks for energy balance, which can prolong the network's lifetime.

We now explain the details of PTCP from the view of sensor i. Sensor i starts to broadcast its priority and location information to neighbors, the sensors within its transmission range. The information is packed as *Initial-Message(i)* which announces the existence of i. At the same time, i will receive the Initial-Messages from its neighbors. We employ multi-hop communication in PTCP so i may receive messages whose sender is out of its transmission range. Sensor i should check the distance between the sender and itself. If the distance is greater than $2D$, the information is abandoned since they are impossible to be in a same circular graph; otherwise, sensor i should record the received information and forward it to neighbors to perform multi hop communication. We set a time window for sensor i to wait for all Initial-Messages. The length of time window l_{win} is determined by sensor deployment density and the range D. It needs to guarantee that all sensors within the range of $2D$ are recorded during the time window. Since information broadcast is usually very fast, the time window should not occupy much time. When the time window ends, sensor i starts to determine whether to sleep. It will broadcast its decision packed as *State-Message(i)*.

There are two kinds of State-Message, (i) State-Message$_{sleep}(i)$, (ii) State-Message $_{active}(i)$, which denote the decision of sensor i respectively. Here we assume C_i contains the recorded sensors from received Initial-Messages and M_i contains sensors whose priority is higher than that of sensor i. If M_i is empty, i occupies the chance to make decision since it is the sensor with the highest priority among the sensors within a distance of $2D$. It will construct the circular graph and divide faces based on the information recorded in C_i. Note that, sensor i itself is not contained in neither C_i nor M_i. Then, it employs Algorithm 3 in Sect. 4.3 to determine whether the region within a distance of D is (D, ϵ)-trap covered. For connectivity issue, sensor i also needs to guarantee that active sensors in a circular graph are connected if it chooses to sleep. Given the transmission range and the location information of all sensors in C_i, i can check whether all sensors in C_i are connected without i. If the region is covered and sensors in C_i are connected without i, sensor i will broadcast a State-Message$_{sleep}(i)$ since it does not need to be active; otherwise, it broadcasts a State-Message$_{active}(i)$ and chooses to stay active. If the region is still not (D, ϵ)-trap covered after sensor i chooses to stay active, it indicates that there are not enough sensors to provide (D, ϵ)-trap coverage and the lifetime of network terminates.

If M_i is not empty, i has to wait the State-Messages from other sensors in M_i. If i receives a new State-Message$_{sleep}$ whose sender is in set C_i, it will record the information, forward the message to its neighbors and remove the sender from set C_i and, M_i if in it; otherwise, it will abandon the message since the message is useless. Sensor i can only make decision when M_i is empty. Then, sensor i constructs circular graph based on C_i. All sensors in C_i are viewed as active when sensor i makes decision. Similarly, sensor i chooses to stay active in PTCP, if probabilistic trap coverage is not guaranteed or sensors in C_i will be disconnected without i. After decision making, sensor i will act as its choice, either in active mode or sleep mode. In summary, PTCP puts sensors into sleep mode in the order of priority/residual energy.

To improve the robustness of our protocol, we set a timeout threshold t_τ in case that State-Messages are dropped accidently during transmission. If M_i is always not empty due to the loss of State-Message, sensor i will wait until the timer exceeds the threshold t_τ. Then it considers that the sensors in M_i choose to be in sleep mode, clears the set M_i and starts to make its own decision. Note that the timer only exceeds the threshold when sensor i receives an Initial-Message(j) but loses the State-Message(j), which is a rare incident since transmission protocols such as TCP/IP can notify the sender if the wireless transmission fails so that the sender can send the message again. We summarize the pseudo-code in Algorithm 4. Also, we illustrate the procedure of PTCP in Fig. 4.9 in the view of sensor i.

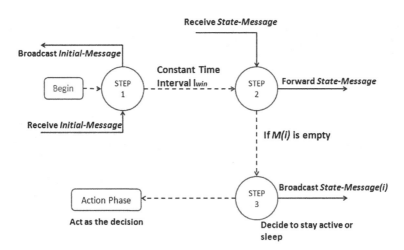

Fig. 4.9 The procedure of sensor i in PTCP

Assume there exist $|V|$ sensors and $|E_d|$ edges in a certain circular graph and F faces in a circle with a radius of $2D$, the time complexity of Algorithm 3 should be $O(F|V||E_d|^3)$ as analyzed in Sect. 4.3.4. Thus, the time complexity of sensor i is also $O(F|V||E_d|^3)$ since Algorithm 3 is the most complex computation part of sensor i during PTCP. It is very efficient compared with a global solution since a circle with a radius of $2D$ is much smaller than a whole region, which leads to a much smaller $|V|$, $|E_d|$ and F. Configuration time is defined as the period during which all sensors have made their decisions. The configuration time of PTCP may be $O(N * F|V||E_d|^3)$ where N is the total amount of sensors in RoI in the worst case since each sensor may have to wait for the response of another sensor. Actually, the timeout threshold t_τ in PTCP should not be less than the configuration time to guarantee that all decisions of sensors in M_i are collected for each sensor i.

For connectivity issue in PTCP, all sensors in a circular graph should be connected as we designed in the protocol. Circular graphs are also connected to each other, otherwise there must exist at least a point between unconnected circular graphs not be probabilistic trap covered. PTCP ensures connectivity while providing probabilistic trap coverage.

4.4.2 Protocol Analysis

We claim that PTCP is an energy-efficient protocol since it minimizes the amount of active sensors and attempts to achieve energy balance among all sensors. PTCP provides an approximation solution to Maximum Trap Network Lifetime Problem. The problem degenerates into the MSC problem if we choose $D = 0$ and $\epsilon = 1$, which has been proven to be NP-complete in [19]. Thus, maximum trap network lifetime problem as a more complex problem is NP-hard. In this section, we analyze the performance of PTCP theoretically and prove the lower bound of the ratio between the lifetime acquired by PTCP and optimum value. Note that we consider a sensor consumes 1 unit of energy if it is active during a complete time slot and the energy consumption of the initial phase of PTCP is neglected since it is rather short compared with a whole time slot. Assume that the lifetime acquired by PTCP is L and the optimum lifetime is L^*. We have Theorem 4.6. E is the amount of time slots that a sensor can be activated, which is usually very large in a real WSN, so the factor $O(1)/E$ is approaching zero. Thus, the lower bound of approximation ratio of PTCP is nearly 1/2.

Theorem 4.6. $L/L^* \geq \frac{1}{2}[1 - \frac{O(1)}{E}]$.

In the following part of this section, we present the proof of Theorem 4.6. We first introduce the concept of *critical sensors* and then prove Lemmas 4.1 and 4.2 which lead to Theorem 4.6.

Firstly, we define *critical sensors* as the minimum amount of sensors without which (D, ϵ)-trap coverage can not be guaranteed. For example, if there exist two disjoint set of sensors and each set of sensors can provide (D, ϵ)-trap coverage, then the amount of critical sensors is at least 2, because we need to remove at least

---

Algorithm 4 Probabilistic trap coverage protocol

1: Define $pr_i = \{E_i, ID_i\}$ as the priority of sensor i. $pr_i > pr_j$ if $E_i < E_j$ or ($E_i == E_j$
   and $ID_i < ID_j$);
2: Set a timeout threshold t_τ.

   % *Find the set of sensors within a distance of* $2D$
3: At the beginning of each time slot, i turns into active mode;
4: Broadcast *Initial-Message(i)* to neighbors;
5: **while** Time window not end **do**
6:    **if** Receive new *Initial-Message(j)* and $distance(i, j) < 2D$ **then**
7:       Record *Initial-Message(j)*;
8:       Broadcast *Initial-Message(j)* to neighbors;
9:    **end if**
10: **end while**

11: Define set C_i contains the recorded sensors from received *Initial-Message*;
12: Define set M_i contains sensors whose priority is greater than pr_i;

   % *Wait for State-Messages and decide whether to stay active or sleep*
13: **while** $C_i \neq \emptyset$ **do**
14:    % *Update* C_i *and* M_i *when receiving State-Message from sensors in* C_i
15:    **if** Receive new *State-Message(j)* **then**
16:       **if** $distance(i, j) < 2D$ **then**
17:          Record *State-Message(j)*;
18:          Broadcast *State-Message(j)* to neighbors;
19:          **if** j decides to sleep **then**
20:             Remove j from C_i;
21:          **end if**
22:          **if** $j \in M_i$ **then**
23:             Remove j from M_i;
24:          **end if**
25:       **end if**
26:    **end if**
27:    % *Clear* M_i *if timeout*
28:    **if** $M_i \neq \emptyset$ and exceed timeout threshold t_τ **then**
29:       $C_i = C_i - M_i$;
30:       Let $M_i = \emptyset$;
31:    **end if**
32:    % *Start to decide if* M_i *is empty*
33:    **if** $M_i == \emptyset$ **then**
34:       Assume set F_i as the faces who are covered by sensor i;
35:       **if** All faces in F_i are (D, ϵ)-trap covered and all sensors in C_i are connected **then**
36:          Broadcast *State-Message$_{sleep}$(i)*;
37:          $mode = sleep$;
38:          break;
39:       **else**
40:          Broadcast *State-Message$_{active}$(i)*;
41:          $mode = active$;
42:          break;
43:       **end if**
44:    **end if**
45: **end while**

---

two sensors so that other sensors can not provide required coverage. The amount of critical sensors, denoted by k, characterizes the degree of redundancy of the network. Usually critical sensors are much less than other sensors. Suppose the set of non-critical sensors is C_{nc}. If any critical sensor is added into C_{nc}, it can provide (D, ϵ)-trap coverage, which is a useful property of critical sensor.

The lifetime of network will terminate if all critical sensors run out of energy. Apparently, we have $L^* \leq kE$. We define probabilistic trap cover C as a set of sensors providing (D, ϵ)-trap coverage to RoI and the coverage requirement can not be met if we remove any sensor in set C. Define set Θ contains all probabilistic trap covers. Each set C in Θ should contain at least one critical sensor according to the definition of critical sensor. Define $\Theta_i = \{C | C \in \Theta, i \in C,$ *no other critical sensors in C*$\}$ for critical sensor i.

The bottleneck of lifetime is the amount of critical sensors because the network can not provide required coverage if all critical sensors run out of energy. The lifetime does not always achieve $k * E$ since there may exist common non-critical sensors when different critical sensors are activated. Consider two critical sensors a and b. If all sets in $\Theta_a \cup \Theta_b$ do not contain any common sensor, a and b can always find a set in Θ_a and Θ_b to activate for E time slots respectively. The lifetime can be at least $2E$. If at least one common sensor i exists in every set in $\Theta_a \cup \Theta_b$, we can not activate a and b separately for E time slots. i is referred to as the *common non-critical sensor*. Common non-critical sensors also restrict the lifetime of network. The critical sensors will consume their energy more quickly than other sensors because at least one critical sensor is needed to be activated each time slot. PTCP activates a critical sensor with the most residual energy each time slot, which makes that all critical sensors have nearly the same residual energy. Common non-critical sensors will almost have the same residual energy as the critical sensors because they have to be activated necessarily with some critical sensors. To discover the property of these sensors, we prove Lemma 4.1.

Lemma 4.1. *Consider two critical sensors a and b. If all sets in $\Theta_a \cup \Theta_b$ contain at least one common sensor i, there must exist a set in Θ containing both a and b without i in it.*

Proof. Suppose there does not exist a set in Θ containing both a and b without i. We adopt reduction to absurdity to prove the lemma. i is not a critical sensor, because neither a nor b can provide (D, ϵ)-trap coverage when added into C_{nc} without sensor i in C_{nc}. Since i is a non-critical sensor, we can reduce the amount of critical sensors by picking i as a critical sensors. Then a and b can be removed from the set of critical sensors. The amount of critical sensors can be reduced, which violates the definition of critical sensors. Thus, there must exist a set in Θ containing both a and b without i in it.

If there exists a set in Θ containing both a and b without i in it, a can still be critical sensor when i is picked as critical sensor since a can provide (D, ϵ)-trap coverage with C_{nc} given $b \in C_{nc}$. Then the amount of critical sensors will not vary.

To find the lower bound of lifetime acquired by PTCP, we prove Lemma 4.2.

Lemma 4.2. $L \geq \frac{1}{2}kE[1 - \frac{O(1)}{E}]$.

Proof. PTCP activates a set of sensors $C \in \Theta$ every time slot. At the beginning, all sensors have the same residual energy. We assume that the critical sensors and common non-critical sensors have less residual energy than any other normal non-critical sensors since time slot α. Since critical sensors and common non-critical sensors have the highest priority to sleep then, there will exist only two cases.

1. No common non-critical sensors have higher priority than their corresponding critical sensors. In this case, only one critical sensor with lowest priority stays active.
2. Some common non-critical sensors have higher priority than their corresponding critical sensors. Without loss of generality, consider common non-critical sensor i which is contained in all sets in $\Theta_a \cup \Theta_b$ and a or b has the lowest priority among all critical sensors. i has higher priority than a and b. There must exist a set containing a and b in Θ according to Lemma 4.1, which is to be activated when i is put into sleep. In this case, two critical sensors will stay active.

In PTCP, critical sensors will always choose to sleep unless probabilistic trap coverage can not be guaranteed due to their relatively low residual energy. We can see that only two critical sensors are needed to be activated at most by PTCP each time slot. Critical sensors consume energy more quickly than others and it takes α time slots to ensure that critical sensors have higher priority. Since at most two critical sensors are needed to be activated each time slot, we have $L \geq \frac{1}{2}k(E-\alpha)$. α has no relevance with E. We therefore denote it as $O(1)$. So we have $L \geq \frac{1}{2}kE[1 - \frac{O(1)}{E}]$, which concludes the proof.

Lemma 4.2 leads to Theorem 4.6 since $L^* \leq kE$, which guarantees the performance of PTCP even in the worst case.

4.5 Performance Evaluation

4.5.1 Environment Setup

In our simulations, we build a WSN with N sensors, each with an initial energy of E units. The sensors are deployed randomly in a square of $10*10$ units^2. The WSN is designed to provide (D, ϵ)-trap coverage. Active sensor in each time slot consumes 1 unit of energy. We assume that the switching frequency is very low so that the communication costs at the beginning of each time slot are negligible in the simulations. We use exponential attenuation probabilistic model introduced in [20] for our problem formulation here.[2] We set $d_1 = 0.2$ units and $d_2 = 0.5$ units in the probabilistic sensing model.

---

[2]The other probabilistic models can also be adopted in a similar way.

There is no existing algorithm designed for probabilistic trap coverage yet, so we employ TCO [2] and RIS [21] to compare with PCTP in the simulations. TCO is proposed to schedule sensors maintaining D-trap coverage with boolean disc sensing assumption. It is the state-of-the-art solution for scheduling in trap coverage. It can also be applied for scheduling in probabilistic trap coverage. We apply it to (D, ϵ)-trap coverage by assigning each sensor a sensing range d_ϵ where d_ϵ is the distance at which the detection probability is the threshold ϵ. Without loss of fairness, the magnitude of sensing range, d_ϵ, is a necessary condition because it can not guarantee (D, ϵ)-trap coverage if its sensing range is less than d_ϵ. RIS is a localized algorithm. There is no communication between sensors at all. Each sensor independently determines to remain awake with a probability p in this time slot. Thus, the coverage performance may not be guaranteed.

At the beginning of each time slot, each sensor employs specified algorithm to determine whether to remain awake. The sensors perform sensing task or sleep during the time slot. Finally, the lifetime of network terminates if (D, ϵ)-trap coverage can not be guaranteed any longer. We achieve this by constructing circular graph for each face and examining the requirement of coverage each time slot. If any point is not (D, ϵ)-trap covered, we will view it as a violation and put an end to the lifetime of network.

The awake probability of RIS is a dilemma. If the probability is very low, the performance will not be guaranteed and the lifetime of network will terminate according to the coverage requirement. The lifetime will also terminate when the energy exhausts quickly if the awake probability is too high, maybe approaching 1. By comparing the performance of different probability, we choose $p = 0.6$ as the representative case with best performance in the following simulations.

4.5.2 Simulation Results

The lifetimes achieved by PTCP, TCO and RIS are plotted in Fig. 4.10. It can be seen that PTCP always has the best performance in various scenarios. As pointed out, energy consumption and energy balance has significantly effects on lifetime of network. PTCP outperforms other approaches due to its better performance in energy consumption and energy balance. Since TCO is not designed for probabilistic sensing model, it can not utilize the detection probability beyond the distance of d_ϵ. Thus it may always employ more sensors than optimal value. RIS is a random algorithm with a fix awake probability p. Thus, the amount of active sensors in RIS is only related with p and the total amount of sensors. It should always be much more than the required value, otherwise the coverage requirement will be violated, which can cause the termination of network lifetime. That is why PTCP is always superior in scheduling sensors to meet the requirements of probabilistic trap coverage. Besides, the lifetime is lengthened if the total amount of sensors N or initial energy E increases. If the requirement of probabilistic trap coverage is relaxed, i.e., D is higher, the lifetime also increases.

Fig. 4.10 Lifetimes of PTCP, TCO and RIS. (**a**) Lifetimes vs. N, $\epsilon = 0.9$, $E = 10$, $D = 2.5$, $V_{max} = 0.1$. (**b**) Lifetimes vs. E, $\epsilon = 0.9$, $N = 140$, $D = 2.5$, $V_{max} = 0.1$. (**c**) Lifetimes vs. D, $\epsilon = 0.9$, $N = 140$, $E = 10$, $V_{max} = 0.1$. (**d**) Lifetimes vs. V_{max}, $\epsilon = 0.9$, $N = 140$, $E = 10$, $D = 2.5$

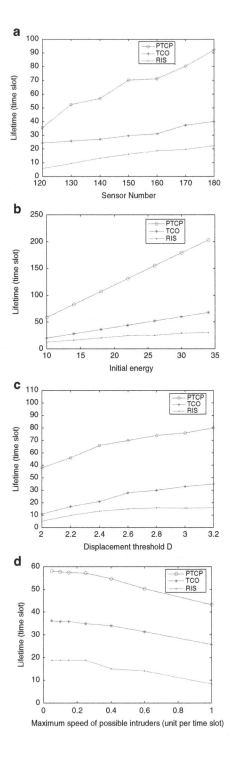

We also compare the lifetimes versus different maximum speed of possible intruders in Fig. 4.10d. The unit of the speed is the amount of units that a target can move within one time slot. If the maximum speed V_{max} decreases, less sensors are needed to perform sensing task, which leads to a longer lifetime. The sensing radius of boolean disc model r of TCO is chosen to ensure that when two sensors located at a distance of $2r$, the detection probability of the target moving across the two sensors along the perpendicular bisector with maximum speed V_{max} is ϵ, so that we can compare these two algorithms fairly. Thus TCO also provides a longer lifetime when the maximum speed decreases due to the increasing of sensing radius. The speed impacts the performance of RIS since the sensing ranges of sensors in RIS also increase which helps cover the region. We may need to deploy more sensors to ensure the longevity of network if the speed of possible intruders is very high.

To evaluate the quality of coverage of PTCP, TCO and RIS, we randomly pick 16 physical points in the RoI and calculate the detection probability of moving target whose origin location is among these points with a displacement of D. By setting each physical point as the start point, we can compute the average detection probability of these sixteen detection probabilities. The average detection probability is calculated during the whole lifetime of network. The average detection probability is plotted in Fig. 4.11. The x-axis is the time slot in the lifetime. The y-axis is the average detection probability of each algorithm in each time slot. We can see that the detection probability of TCO and RIS are much higher than the threshold, which indicates that they activate much more sensors than the required degree. It is a waste of energy anyway. Since the detection probabilities of TCO and RIS are both approaching 1, so they are difficult to differ. PTCP utilizes the margin between actual detection probability and the threshold ϵ by constructing the circular graph so that it can activate less sensors than other algorithms.

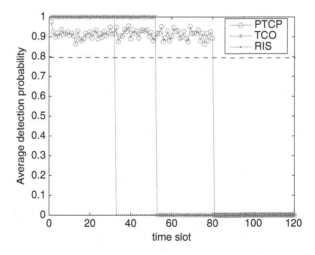

Fig. 4.11 Detection probability of moving target, $E = 20$, $N = 150$, ϵ=0.8, D=2.5, V_{max}=0.1

4.6 Conclusion

In this chapter, we have analyzed the detection probability of a moving target theoretically based on the probabilistic sensing model and developed the theory of circular graph. We successfully find a lower bound of detection probabilities among all possible paths based on the theory of circular graph. Besides, the theory of circular graph can also be applied in other areas of sensing coverage such as barrier coverage. Actually we have applied the circular graph to solve a problem of closed-belt barrier coverage in this chapter.

Based on the analysis of detection probability, we introduced the concept of probabilistic trap coverage and formulated the maximum trap network lifetime problem. We proposed the Circular Graph Algorithm to check whether a given sensor network can provide required probabilistic trap coverage. A localized protocol PTCP was proposed to schedule the activation of sensors with the help of circular graph algorithm and it was observed by simulations to be much more efficient than the state-of-the-art solutions. We also analyzed the performance of PTCP theoretically and established a lower bound of lifetime acquired by the protocol to be nearly 1/2 of the optimum lifetime, which suggests that the protocol can always perform well even in the worst case.

References

1. P. Balister, Z. Zheng, S. Kumar, and P. Sinha. Trap coverage: Allowing coverage holes of bounded diameter in wireless sensor networks. In *Proceedings of IEEE International Conference on Computer Communications (INFOCOM)*, 2009.
2. J. Li, J. Chen, S. He, T. He, Y. Gu, and Y. Sun. On energy-efficient trap coverage in wireless sensor networks. In *Proceedings of IEEE Real-Time Systems Symposium (RTSS)*, 2011.
3. R. Tan, G. Xing, B. Liu, and J. Wang. Impact of data fusion on real-time detection in sensor networks. In *Proceedings of IEEE Real-Time Systems Symposium (RTSS)*, 2009.
4. G. Xing, R. Tan, B. Liu, J. Wang, X. Jia, and C. Yi. Data fusion improves the coverage of wireless sensor networks. In *Proceedings of the Annual International Conference on Mobile Computing and Networking (MobiCom)*, 2009.
5. I. Altinel, N. Aras, E. Güney, and C. Ersoy. Binary integer programming formulation and heuristics for differentiated coverage in heterogeneous sensor networks. *Computer Networks*, 52(12):2419–2431, 2008.
6. J. Chen, J. Li, S. He, Y. Sun, and H. Chen. Energy-efficient coverage based on probabilistic sensing model in wireless sensor networks. *IEEE Communication Letters*, 14(9):833–835, 2010.
7. N. Ahmed, S. Kanhere, and S. Jha. Probabilistic coverage in wireless sensor networks. In *Proceedings of the IEEE Conference on Local Computer Networks 30th Anniversary (LCN)*, 2005.
8. S. Kumar, T. Lai, and A. Arora. Barrier coverage with wireless sensors. In *Proceedings of the Annual International Conference on Mobile Computing and Networking (MobiCom)*, 2005.
9. J. Chen, J. Li, and T. H. Lai. Trapping mobile targets in wireless sensor networks: An energy-efficient perspective. *IEEE Transactions on Vehicular Technology*, 62(7):3287–3300, 2013.

10. G. Mao, B. Fidan, and B. Anderson. Wireless sensor network localization techniques. *Computer Networks*, 51(10):2529–2553, 2007.
11. N. Patwari, J. Ash, S. Kyperountas, A. Hero, R. Moses, and N. Correal. Locating the nodes: cooperative localization in wireless sensor networks. *IEEE Signal Processing Magazine*, 22(4):54–69, 2005.
12. Y. Zou and K. Chakrabarty. Sensor deployment and target localization in distributed sensor networks. *ACM Transactions on Embedded Computing Systems*, 3(1):61–91, 2004.
13. S. Kumar, T. Lai, M. Posner, and P. Sinha. Maximizing the lifetime of a barrier of wireless sensors. *IEEE Transactions on Mobile Computing*, 9(8):1161–1172, 2010.
14. J. Li, J. Chen, and T. Lai. Energy-efficient intrusion detection with a barrier of probabilistic sensors. In *Proceedings of IEEE Conference on Computer Communications (INFOCOM)*, 2012.
15. D.B. West. *Introduction to Graph Theory*. Prentice Hall, 2001.
16. T. Cormen, C. Leiserson, R. Rivest, and C. Stein. Data structures for disjoint sets. In *Introduction to Algorithms*, pages 498–524. MIT Press and McGraw-Hill, 2001.
17. B. Liu, O. Dousse, J. Wang, and A. Saipulla. Strong barrier coverage of wireless sensor networks. In *Proceedings of the ACM International Symposium on Mobile Ad Hoc Networking and Computing (MobiHoc)*, 2008.
18. A. Chen, S. Kumar, and T. Lai. Designing localized algorithms for barrier coverage. In *Proceedings of the Annual ACM International Conference on Mobile Computing and Networking (MobiCom)*, 2007.
19. M. Cardei, T. Thai, Y. Li, and W. Wu. Energy-efficient target coverage in wireless sensor networks. In *Proceedings of IEEE International Conference on Computer Communications (INFOCOM)*, 2005.
20. Y. Zou and K. Chakrabarty. A distributed coverage and connectivity centric technique for selecting active nodes in wireless sensor networks. *IEEE Transactions on Computers*, 25(8):978–991, 2005.
21. S. Kumar, T. Lai, and J. Balogh. On k-coverage in a mostly sleeping sensor network. In *Proceedings of the Annual International Conference on Mobile Computing and Networking (MobiCom)*, 2004.

Chapter 5
Conclusions

Last decade has witnessed the rapid advance in the *energy-efficient area coverage* in sensor networks. Most of existing literature focused on disc sensing model and tried to design dynamic duty-cycle scheduling algorithms to improve energy-efficiency while maintaining *area coverage* performance. These work provided a desirable solution to general area coverage problem and a basis to further investigation. Based on these existing work, it becomes pressing to explore new approaches to further improve the *energy-efficient area coverage*. Apparently, coverage performance closely hinges on the unique requirements of applications, and the intrinsic characteristics of the applications should be exploited. We in this book illustrate how to explore the potential of improving *energy-efficient area coverage* by focusing on the intruder detection applications.

In Chap. 1, we first introduced the background of sensor networks and the implication of coverage. We also elaborated on system model such as the formal definition of area coverage and sensing models in Chap. 1 to help readers understand the book easily. We gave a clear introduction to existing literatures and applications on *area coverage*. Then we focused on energy-efficient intruder detection under the well-known binary sensing model in Chap. 2. In this chapter, we showed how to improve the energy efficiency without impairing coverage performance by exploiting the dynamic nature of stochastic intruders. In Chap. 3, we proceeded to investigate the intruder trapping under binary sensing model. By intruder trapping, we refer to the sensing requirement that any moving intruder will be detected by sensor networks when its movement distance is greater than a predefined constant. The network coverage holes act as traps. Once intruders fall into these traps, it can not get out without being detected. This scenario is interesting when we try to capture the moving intruders in the surveillance region. We designed efficient scheduling algorithms to prolong the network lifetime while guaranteeing the intruder trapping performance. To investigate the impact of different sensing models on the performance intruder trapping, we studied the problem under the probabilistic sensing model in Chap. 4. We hope that good approaches can be inspired to yield salient solutions to the problems in other applications.

S. He et al., *Energy-Efficient Area Coverage for Intruder Detection in Sensor Networks*, SpringerBriefs in Computer Science, DOI 10.1007/978-3-319-04648-8_5,